Indigenous Heritage

History shows that travellers sought to experience the unfamiliar and exotic cultures and traditions of Indigenous peoples, with early examples of Indigenous tourism in the United States, Canada, Scandinavia, Australia, New Zealand and countries throughout Asia and Latin America. Similarly, contemporary travellers demonstrate a desire to seek out opportunities to experience Indigenous peoples and their cultures. Thus, we are witnessing worldwide growth in the awareness of, and interest in, Indigenous cultures, traditions, histories and knowledges.

Engagement in the tourism sector is regularly advocated for Indigenous peoples because of the socio-economic opportunities it provides; however, there are a range of cultural benefits including the maintenance, rejuvenation and/or preservation of Indigenous cultures, knowledges and traditions for Indigenous peoples who choose tourism as a vehicle to showcase their cultures. Consequently, tourism is regularly acknowledged as a means for facilitating the sustainability of tangible and intangible Indigenous cultural heritage including languages, stories, art, dance, rituals and customs. Importantly, however, the history of Indigenous peoples' engagement in tourism has provided a range of examples of the threats to Indigenous culture that can accrue as a result of tourism (i.e., cultural degradation, commercialisation and commodification, authenticity and identity, among others). This book presents an exploration of the intersection between tourism and Indigenous culture.

The chapters in this book were originally published as a special issue of the *Journal of Heritage Tourism.*

Michelle Whitford is Associate Professor and Dean (Learning and Student Outcomes) in the Office of the Senior Deputy Vice Chancellor at Griffith University. Her research expertise is in the field of policy, planning, sustainable development and management of Indigenous tourism and events. Whitford's expertise includes co-coordinating projects in the area of Indigenous tourism and events with a focus on supply and demand, capacity development, entrepreneurship, authenticity and commodification and management. She has co-coordinated research projects for various organisations including the Sustainable Tourism Cooperative Research Centre, the Australian Institute of Aboriginal and Torres Strait Island Studies, Indigenous Business Australia, National Environmental Science Program and numerous Australian Federal State and Local Government agencies.

Lisa Ruhanen is Professor in Tourism and the Director of Education for the UQ Business School, The University of Queensland. She has undertaken more than 30 academic and consultancy research projects in Australia and overseas in the areas of Indigenous tourism, sustainable tourism and policy, planning and governance. Lisa has more than 100 academic publications and in 2017 she and colleagues co-edited a book on *Indigenous Tourism: Cases from Australia and New Zealand.* Lisa teaches undergraduate and postgraduate courses in sustainable and responsible tourism, ethics and tourism in developing countries. She has worked closely with the United Nations World Tourism Organization over the last decade and is currently an Advisory Board member and auditor for the UNWTO's TedQual accreditation program.

Indigenous Heritage

Edited by
Michelle Whitford and Lisa Ruhanen

Routledge
Taylor & Francis Group
LONDON AND NEW YORK

First published 2022
by Routledge
2 Park Square, Milton Park, Abingdon, Oxon OX14 4RN

and by Routledge
605 Third Avenue, New York, NY 10158

Routledge is an imprint of the Taylor & Francis Group, an informa business

British Library Cataloguing in Publication Data
A catalogue record for this book is available from the British Library

ISBN: 978-0-367-69721-1 (hbk)
ISBN: 978-0-367-69723-5 (pbk)
ISBN: 978-1-003-14299-7 (ebk)

Typeset in Minion Pro
by Newgen Publishing UK

Publisher's Note
The publisher accepts responsibility for any inconsistencies that may have arisen during the conversion of this book from journal articles to book chapters, namely the inclusion of journal terminology.

Contents

Citation Information

The chapters in this book were originally published in the *Journal of Heritage Tourism*, volume 14, issue 3 (2019). When citing this material, please use the original page numbering for each article, as follows:

Chapter 1

Cultural heritage and Indigenous tourism
Lisa Ruhanen and Michelle Whitford
Journal of Heritage Tourism, volume 14, issue 3 (2019), pp. 179–191

Chapter 2

Evolution of indigenous tourism among the Lacandon of Chiapas: an application of Weaver's model
Pilar Espeso-Molinero and María José Pastor-Alfonso
Journal of Heritage Tourism, volume 14, issue 3 (2019), pp. 192–204

Chapter 3

Cultural sustainability – a framework for Aboriginal tourism in British Columbia
Tatjana Thimm
Journal of Heritage Tourism, volume 14, issue 3 (2019), pp. 205–218

Chapter 4

Developing indigenous tourism in the bomas*: critiquing issues from within the Maasai community in Tanzania*
Kokel Melubo and Anna Carr
Journal of Heritage Tourism, volume 14, issue 3 (2019), pp. 219–232

Chapter 5

Settler colonialism, Indigenous cultures, and the promotional landscape of tourism in Ontario, Canada's 'near North'
Bryan S. R. Grimwood, Meghan L. Muldoon and Zachary M. Stevens
Journal of Heritage Tourism, volume 14, issue 3 (2019), pp. 233–248

Chapter 6

Authenticity as a compromise: a critical discourse analysis of Sámi tourism websites
Cecilia de Bernardi
Journal of Heritage Tourism, volume 14, issue 3 (2019), pp. 249–262

Chapter 7

Indigenous tourism in Australia: understanding the link between cultural heritage and intention to participate using the means-end chain theory
Trinidad Espinosa Abascal
Journal of Heritage Tourism, volume 14, issue 3 (2019), pp. 263–281

Chapter 8

The interpreter as researcher: ethical heritage interpretation in Indigenous contexts
Chance Finegan
Journal of Heritage Tourism, volume 14, issue 3 (2019), pp. 282–294

Notes on Contributors

Anna Carr, Department of Tourism, University of Otago, Dunedin, New Zealand.

Cecilia de Bernardi, Faculty of Social Sciences, Multidimensional Tourism Institute, University of Lapland, Rovaniemi, Finland.

Pilar Espeso-Molinero, Social Anthropology Department, University of Alicante, Alicante, Spain.

Trinidad Espinosa Abascal, College of Business, Victoria University, Melbourne, VIC, Australia.

Chance Finegan, Faculty of Environmental Studies, York University, Toronto, Canada.

Bryan S. R. Grimwood, Department of Recreation and Leisure Studies, University of Waterloo, Waterloo, ON, Canada.

Kokel Melubo, College of African Wildlife Management, Mweka, Tanzania.

Meghan L. Muldoon, Department of Recreation and Leisure Studies, University of Waterloo, Waterloo, ON, Canada.

María José Pastor-Alfonso, Social Anthropology Department, University of Alicante, Alicante, Spain.

Lisa Ruhanen, UQ Business School, The University of Queensland, Brisbane, Australia.

Zachary M. Stevens, Department of Recreation and Leisure Studies, University of Waterloo, Waterloo, ON, Canada.

Tatjana Thimm, Department of Business, Cultural and Legal Studies, HTWG Konstanz, Konstanz, Germany.

Michelle Whitford, Griffith Business School, Griffith University, Brisbane, Australia.

Cultural heritage and Indigenous tourism

Lisa Ruhanen and Michelle Whitford

ABSTRACT

Tourism is acknowledged as a vehicle that can help sustain both tangible and intangible elements of Indigenous cultural heritage, including languages, stories, song, art, dance, hunting methods, rituals and customs. Often, cultural heritage products developed for tourism promise to provide many socio-economic opportunities for the communities involved, however, tourism can also present a challenge as the self-management of Indigenous cultural product and cultural identity can be problematic. Given the pivotal role culture plays in the sustainability of Indigenous tourism products, it is time for a twenty-first century examination of the nexus between cultural heritage and Indigenous tourism. This paper reflects upon tourism and contemporary Indigenous cultural heritage, both tangible and intangible. Through a review of the current literature on Indigenous tourism and cultural heritage, the authors identify key areas for future research and aim to stimulate further discussion around the ways Indigenous tourism may be developed to sustain Indigenous cultural heritage.

Introduction

The worldwide Indigenous population is estimated to be around 370 million (United Nations, 2017). Indigenous peoples, who represent a significant part of the world's vast cultural and linguistic diversity and heritage, possess unique knowledge systems and are the keepers of most of the world's intangible cultural heritage (The World Bank, 2018; United Nations, 2017; Warnholtz & Barkin, 2017). Indigenous peoples make up only five percent of the global population yet they account for about one-third of the global poor and fifteen percent of the extreme poor (The World Bank, 2018; United Nations, 2017). Additionally, an estimated 65% of the world's land is under Indigenous customary ownership (Rights and Resources Initiative, 2015); however, many governments around the globe only recognise a fraction of land as formally or legally belonging to Indigenous peoples (The World Bank, 2018). Further, the social, economic and political marginalisation of Indigenous peoples remains pervasive in all the regions of the world even though Indigenous knowledge systems are recognised as crucial for ongoing sustainable development (United Nations, 2017),.

As governments continue to focus on improving the socio economic disadvantage experienced by too many Indigenous peoples, tourism has long been advocated as a social intervention that can be used to address a range of social, economic and political issues and as a possible trigger for the development of local host communities (de Kadt, 1979a, 1979b). The possibility of attracting investment, foreign currency, creating jobs and eventually creating the opportunity for economic growth for host communities has also led to the incorporation of tourism into the poverty reduction, sustainable development and social enterprise agendas, especially for communities with strong natural and cultural assets, both intangible and tangible (Korstanje, 2012; Warnholtz & Barkin, 2018). For those Indigenous peoples who choose to share their culture through touristic endeavours, tourism is often lauded for its potential to provide a range of benefits including the maintenance, rejuvenation and/or preservation of Indigenous cultures, knowledges and traditions. Tourism has also been acknowledged as a vehicle that can help sustain intangible elements of Indigenous cultural heritage, including languages, stories, song, art, dance, hunting methods, rituals and customs (Burns, 2006; Einar Johansen & Mehmetoglu, 2011; Warnholtz & Barkin, 2018; Whitney-Squire, 2016).

As a growing number of Indigenous communities around the world engage with tourism as a vehicle for reforming their economies, consideration needs to be given, not only to the optimistic predictions regarding the benefits of tourism but also to the negative impacts tourism can inflict upon communities' tangible and intangible cultural assets. Indeed, the history of Indigenous peoples' engagement in tourism provides numerous examples of the threats to Indigenous culture that can accrue as a result of tourism. For instance, cultural degradation, commercialisation and commodification of intangible cultural heritage and loss of identity are but some of the ways tourism continues to negatively affect Indigenous cultures (McIntosh, Hinch, & Ingram, 2002; Ranasinghe & Cheng, 2018). Consequently, threats to intangible cultural heritage are problematic, not only for the maintenance and preservation of the culture but also for the commercial sustainability of the 'cultural product' and any long-term socio-economic benefits tourism may bring to a community and/or a region, yet they are increasingly being seen as unique tourism 'assets' (Nielsen & Wilson, 2012).

Thus, the increasing propensity for governments around the globe to utilise Indigenous tourism 'assets' as a socio-economic growth strategy for Indigenous communities and/or regions has given rise to an increasing number of academics focusing their research in this field. For example, recent overviews of this research (for example: Carr, Ruhanen, & Whitford, 2016; Nielsen & Wilson, 2012; Whitford & Ruhanen, 2016) revealed that while a wide range of topics are addressed in the literature (i.e. social enterprise and entrepreneurship, employment, self-determination), there is arguably a clear interest in the socio-economic agenda of development that is reflective of the wider development agenda.

Given the pivotal role culture plays in the development of sustainable Indigenous cultural tourism products, and the increasing significance governments are placing on the development of Indigenous tourism, an examination of the nexus between cultural heritage and Indigenous tourism is timely. As the opening paper of this special issue on cultural heritage and Indigenous tourism, this paper reflects upon contemporary Indigenous cultural heritage and tourism, both tangible and intangible. By highlighting key themes within the literature, this paper seeks to identify key areas for future research and stimulate further discussions around the ways Indigenous tourism may be developed to sustain Indigenous cultural heritage.

Labelling tourism as 'Indigenous'

Historically, the term Indigenous peoples was used to refer to descendants of those who occupied a given territory that was invaded, conquered or colonised by white colonial powers (Saugestad, 2000; Weaver, 2016) and as such, the object (and increasingly the subject) of Indigenous tourism has widely come to be seen as consisting of a group of culturally differentiated people who lived and occupied territories before the existence of the nation state (Saarinen, 2013). More recently, however, there has been a process of inclusion of Africans in the global Indigenous movement (Giblin, 2017; Kagumba, 2013; Saugestad, 2008). In some cases, such as the Maasai in Kenya, the labelling of African groups as 'Indigenous' has been undertaken in an attempt to access Indigenous tourism markets established elsewhere, such as in Australia, New Zealand, USA and Canada. In other cases, such as the Batwa in Uganda, it has been done to align with the Indigenous rights movement to protest, for example, eviction from cultural territories (Giblin, 2017). Interestingly, the use of the terms 'Indigenous' and 'Indigenous tourism' in the African context, and the extent to which these terms are applicable and relevant, is a topic of much debate, not only amongst scholars and researchers but also between African governments (Kagumba, 2013).

Such debate, however, is not a new phenomenon. Since the emergence of Indigenous tourism in the 1970s, there has been a variety of definitions pertaining to what constitutes Indigenous tourism (for an overview of these definitions see Pereiro, 2016). However, fundamental to these definitions is the centrality of Indigenous culture and identity. Many researchers also agree that a second crucial definitional element is the locus of control (Hinch & Butler, 2007; Pereiro, 2016; Zeppel, 2010). That is, Indigenous peoples should not be the focus of the tourism attractions/products per se, but rather tourism attractions/products should be 'based on the group's land and cultural identity and controlled from within the group'. Thus, the expectation is that Indigenous peoples will have control of the way their culture (tangible and intangible) is portrayed and accessed through tourism endeavours. It has been suggested, however, that in reality, Indigenous tourism actually works to uphold neo-colonialist attitudes (Korstanje, 2012) resulting in too many Indigenous communities/regions still being subjected to ethnocentric treatment, which includes 'protection' (from tourist consumption), by outsiders who believe they know better than Indigenous peoples (Richards, 2018). Indeed, in too many instances, Indigenous tourism ventures are developed according to the needs and priorities of non-Indigenous people (Nielsen & Wilson, 2012; Williams & Gonzalez, 2017), who often forget or neglect the very people(s) from which the tourism attractions/products are constructed (Tomaselli, 2012). Therefore, if the goal is to ensure Indigenous tourism cultural products develop sustainably, as we increasingly draw upon Indigenous culture to develop Indigenous tourism attractions/product, at the very least, we must seek to gain a more wholistic understanding of the role Indigenous cultural identity and cultural representation plays in the sustainable development of cultural Indigenous tourism products.

Identity, authenticity and the commercialisation of Indigenous culture

Tourism has the capacity to shape/or transform the culture of an Indigenous community (Ranasinghe & Cheng, 2018) by influencing the way individuals in a community see themselves and how they perceive their identity and in particular, their intangible cultural heritage (Heldt Cassel & Maureira, 2017). For instance, as a result of tourism, cultural practices and traditions are increasingly transformed into tourism products (Maraud & Guyot, 2016; Ryan & Aiken, 2015) and in the process, Indigenous identity is often reconstructed to satisfy tourist expectations (Coronado, 2014; Heldt Cassel & Maureira, 2017; Pereiro, 2016; Ryan & Aiken, 2015). Thus, while tourism presents an array of opportunities for Indigenous communities to increase awareness and appreciation of, their intangible cultural heritage, concomitantly tourism poses a challenge for those same communities to preserve and maintain the integrity of their intangible cultural heritage and associated 'tourism assets' (Ranasinghe & Cheng, 2018). Arguably, a cost benefit analysis is required in this instance to determine if the gain outweighs the cost.

For instance, due to modernisation, changes in lifestyle, livelihood and spiritual inclination, some Indigenous peoples have apparently lost connection with their culture and their identities (Ona & Solis, 2017). Tourism practices, however, can play an essential role in providing opportunities for Indigenous peoples (and others) to reconnect, discover, rediscover and/or reinforce their identities through engagement with their cultural heritage (i.e. traditional dance, crafts and art) (Burns, 2006; Espinosa Abascal, Fluker, & Jiang, 2015; Stronza, 2008). Involvement in tourism can also provide individuals with opportunities to learn Indigenous languages and consider or reinforce the importance of belonging to an Indigenous community (Heldt Cassel & Maureira, 2017). Moreover, tourism can be used to help support the development, revitalisation and/or preservation of vulnerable and critically endangered languages (Whitney-Squire, 2016). Indeed, the progressive loss of Indigenous languages globally is a serious threat to sustaining Indigenous peoples' identities and intangible cultural heritage.

Similarly, income generating activities such as handicrafts and food security (i.e. traditional activities such as hunting) present a range of opportunities for Indigenous communities to develop entrepreneurial and/or social enterprise tourism ventures (Ona & Solis, 2017). In Australia, for example, modern Aboriginal art is an intrinsic part of a cultural revival. Community art centres, owned by Aboriginal communities and which are largely not-for-profit organisations, have a mandate that includes the promotion of cultural sustainability and the sharing of culture with visiting tourists (Butler, 2017).

We can see then, that tourism can act as a platform in which identities and culture can be contested and/or developed (King & Borges de Lima, 2018). Culture and identity are recognised as dynamic rather than static (Cohen, 1988), thus tourism can be used by Indigenous communities as a platform to revitalise, define, shape and transform their culture within a contemporary context and as 'a means of transmitting their culture, values and beliefs, and their ways of relating to the world of visitors' (Campos, Scott, & Breaky, 2014, p. 143). Resultant changes in cultural practice or the recontextualising of culture for socio-economic purposes are not necessarily indicative of cultural loss and/or loss of meaning for the cultural product or cultural symbols (Kagumba, 2013). Instead, such changes can be viewed as cultural progression and a change in meaning.

It has also been argued that the commoditisation of Indigenous culture can be viewed as a form of empowerment for the Indigenous peoples involved as the culture that was once a source of marginalisation instead comes to hold socio-economic value and becomes celebrated and treasured in tourism contexts (Kagumba, 2013; Pookhao, Bushell, Hawkins, & Staiff, 2018). Indeed, several authors have emphasised the important role tourism plays in enabling Indigenous hosts to perform chosen aspects of their identities that benefit contemporary life within the communities and reflect a combination of tourism and local imaginaries (Bunten, 2008, 2014; Heldt Cassel & Maureira, 2017; Kagumba, 2013; Mackley-Crump, 2016; Travesi, 2018). This gives the hosts the power to act as cultural brokers and the capacity to subvert dominant paradigms and tourists' representations and expectations about indigeneity (Bunten, 2014; Travesi, 2018).

While the process of establishing local Indigenous identities and cultures in the global tourist market can result in the reinvention, recreation and preservation of Indigenous cultural attributes, it may, however, serve to impact on the privacy of the Indigenous peoples involved and result in a sense of invasion and threat (Pereiro, 2016). Thus, when Indigenous tourism is no longer an organised, commercial activity/product and instead becomes a 'way of community life', we must question the extent to which Indigenous tourism threatens Indigenous identity and culture. For example, in a study which focused on the effect of tourism on the Indigenous Vedda community in Sri Lanka, Ranasinghe and Cheng (2018) found that while Indigenous cultural tourism resulted in the opening up of this remote, culturally distinctive space to global networks, the community and their culture and traditions have been pushed to the wayside by new regulations and neoliberal economic conditions. Critical modifications to the Vedda culture and traditional way of life are threatening the survival of the Vedda people's cultural heritage, identity and language and as a result, the values of the Vedda have altered as they have become a money-driven, commodity-oriented group leading a hybrid life: one real and one for the sake of tourists' amusement.

This hybrid existence, forged by Indigenous peoples to cope with tourism, is well documented in the literature (Bresner, 2014; Bunten, 2014; Coronado, 2014; Heldt Cassel & Maureira, 2017; Liu, 2018; McIntosh et al., 2002; Pereiro, 2016; Ranasinghe & Cheng, 2018; Stronza, 2008). For Indigenous hosts, expressing a visual, Indigenous identity in tourism is often a balancing act between maintaining a level of recognition and familiarity that mirrors the expectations of the public imagination and at the same time, conveying a representation that is locally meaningful and emblematic to the Indigenous hosts (Bresner, 2014). It also represents a paradox. There is an expectation that the Indigenous culture that has been commercialised as a tourism product is one that has been preserved from the pre-modern era and therefore considered authentic (at least in terms of tourist expectations), yet in such a discourse, the complexities of culture, from the Indigenous hosts' perspectives, are largely excluded (Coronado, 2014). This means, due to their economic dependence on tourists, Indigenous hosts typically accommodate guests' desires rather than demanding that guests accommodate themselves to the situation of the tour destination (Hsu & Nilep, 2018). At the same time, when tourists value the authenticity of archaic or 'traditional' practices, destination areas and their

Indigenous inhabitants are not able to evolve in response to changing economic or historical situations (Hsu & Nilep, 2018) and, when an Indigenous culture is transformed as a result of its commercialisation for tourism, it often comes to be viewed by tourists as inauthentic (Bunten, 2008; Coronado, 2014; Mackley-Crump, 2016; Ruhanen, Whitford, & McLennan, 2015; Travesi, 2018).

As tourists' perceptions of authenticity are an essential component in Indigenous tourism (Hsu, 2018; McIntosh et al., 2002), questions regarding how to maintain the authenticity of Indigenous culture, while at the same time allowing an Indigenous group to evolve in response to changing socio-economic and cultural situations are central in the quest to ensure the sustainable development of the Indigenous tourism product. To thus contribute to the analysis of authenticity in a manner that supports sustainable development by considering host perspectives, a growing number of authors are conceptualising authenticity within the dynamic context of living culture in Indigenous tourism destinations (for example see: Bresner, 2014; Bunten, 2008, 2014; Heldt Cassel & Maureira, 2017; Hsu, 2018; Hsu & Nilep, 2018; Kagumba, 2013; Mackley-Crump, 2016; Ranasinghe & Cheng, 2018; Travesi, 2018). Such research increasingly emphasises the amorphous nature of 'authentic' cultural practices, recognising them as constantly 'in transition' (Bresner, 2014, p. 138), negotiated and reinvented in relation to both host–guest relationships and resultant of the complicated crossovers and cultural mixings that are increasingly common in a globalised world. Authenticity in this context is thus seen as being dependent on how actors 'choose to authentically represent themselves' (Mackley-Crump, 2016, p. 173). These approaches also attempt to shift the tourist gaze from seeking something predetermined and 'frozen in time' to instead viewing culture and cultural identity as something more representative of the dynamic interaction between modernity and tradition.

Yet, despite moves to understand and view Indigenous cultural heritage from a more dynamic host-centred perspective, these analyses of 'interactive, dynamic' authenticity generally do not probe the power dimensions, 'othering' and forms of exploitation that are at play when tourists gaze upon the everyday lives of Indigenous peoples or their tropes in promotional literature (Bott, 2018). This means hegemony continues to be perpetuated through popular representations and the commercialisation of Indigenous cultures (Richards, 2018), and too often the culture and identities of Indigenous peoples continue to be reduced into a commodity to be bought and sold (Coronado, 2014). Around the world, many Indigenous people still find themselves with very little control of the imagery that commodifies their difference by stereotyping them as, for example, 'exotic', 'mysterious' and 'timelessly authentic', and which thus establishes tourist expectations regarding what constitutes an Indigenous tourism experience, as well as the authenticity of that experience. Indigenous peoples, therefore, continue to find themselves in the position of the 'powerless' where their culture is externally commercialised and the sharing of their culture is neither a consultative process, nor a process that considers Indigenous rights. For example, Liu (2018) shows how the process of developing a cultural heritage location in Southwest China into a tourism destination that is built to satisfy the touristic 'alien imagination' (p. 362) rather than Indigenous viewpoints, has led to the cultural standardisation and delocalisation of the location and thus the erasure of distinctive locality. In another example, Williams and Gonzalez (2017) show how, in the creation of Hawaii's unique tourism niche, native Hawaiian culture has been commodified in a way that continues to be invasive, unsustainable and ultimately, incompatible with Indigenous sovereignty. The authors show how tourism continues to be used to smooth over issues related to colonial occupation, while the Indigenous peoples are left to grapple with what it means to work in a form of Indigenous tourism within this context of ongoing occupation. Bott (2018) also explores how minority Indigenous groups in Vietnam continue to be represented in ways that reproduce certain racial and gendered tropes and how the power plays from this 'othering' has considerable and troubling effects on Indigenous groups. In his review of cultural research trends, Richards (2018, p. 16) provides several examples where minority voices in Indigenous cultural tourism are 'currently drowned out by dominant narratives and ignored by Western analyses of the tourism system', thus suggesting there is a need to listen more carefully to minority voices.

These examples indeed highlight the complexity of issues surrounding identity, authenticity, and the commercialisation of Indigenous cultural heritage. They also show that, despite global efforts to increase Indigenous voices and control in tourism (Nielsen & Wilson, 2012), questions of cultural ownership and the 'right' to share culture still, too often neglect the very people whose culture is being shared. From a sustainability perspective, as long as questions concerning ownership, control, authenticity and the commercialisation of Indigenous cultural heritage remain an ongoing issue, the ability for Indigenous communities involved in tourism to develop in sustainable ways also remains questionable.

Sustainable development of Indigenous cultural heritage through tourism

Sustainable development and sustainability are recognised as important and essential objectives in the endeavour to foster Indigenous tourism enterprises that will achieve long-term success and which will contribute to the socio-economic and cultural improvement of Indigenous lives. Development, sustainability and the combination thereof, however, have been heavily contested concepts (Warnholtz & Barkin, 2018) that largely stem from a Western definition of development. In the contemporary definition of sustainable development, development has underpinning values of the neoliberal mainstream market, relative equity, self-determination and self-empowerment. These are terms that are all derived from prioritising individual well-being and interests over those of the community (Warnholtz & Barkin, 2018). From this perspective, Indigenous tourism ventures are considered successful if Indigenous communities are involved in and or have some control over them (Kessy, 2017). Indeed, academics have long stressed the importance of developing tourism in harmony with Indigenous community interests and in a way that ensures Indigenous people maintain ownership and control over product development and cultural experiences (for example: Altman & Finlayson, 1992; King & Stewart, 1996; McIntosh et al., 2002). The ongoing failure of many tourism programmes as 'sustainable development tools', however, (Carr et al., 2016; Ruhanen & Whitford, 2018) suggests that benchmarking success on some form of involvement or control by Indigenous communities is simply not appropriate, nor sustainable. In many cases the failures are not caused by lack of involvement or control, but instead by the language of the objectives and the policies of development agencies which prioritise the sustainability of the tourism sector over that of the host communities' culture (Warnholtz & Barkin, 2018).

Given the ongoing problems arising from Western-centric attempts to 'sustainably' develop Indigenous enterprises for tourism, critical voices are increasingly proposing alternative ways to development and sustainability that take a much greater community-driven approach. Scholars have sought to shift the emphasis away from the immediate needs of visitors and the tourism industry in general to instead consider the cultural values at stake within tourism development (for example: Carr et al., 2016; Pereiro, 2016; Saarinen, 2013; Scherrer & Doohan, 2013; Thompson-Carr, 2013; Warnholtz & Barkin, 2018; Whitford & Ruhanen, 2016; Whitney-Squire, 2016; Zeppel, 2010). Foundational to this shift is the consideration of Indigenous peoples' relationship with the land and by extension, the cultural space in which tourism initiatives can support sustainable, meaningful engagement within the industry (Whitney-Squire, 2016, p. 1169). For an Indigenous enterprise to be sustainable, the Indigenous groups involved need not only be given control of the tourism enterprises in which their culture is the central element, they also need to be given the capacity to build their own future, as well as the ability to exert political control over their cultural and natural resources (Carr et al., 2016; Espeso-Molinero, Carlisle, & Pastor-Alfonso, 2016; Kessy, 2017; Warnholtz & Barkin, 2018). Projects thus need to be developed in a manner that corresponds to a community's own values and future aspirations, allowing them to draw on lessons of their historical experience, as well as of their real and potential cultural resources (Warnholtz & Barkin, 2018; Yang, 2012). Indeed, embedding Indigenous values and world-views in tourism development is increasingly being emphasised as unequivocally necessary to affect positive outcomes (Carr et al., 2016).

One alternative approach to development that delivers a model embedded in Indigenous values and world-views is ethnodevelopment (see: Borges de Lima & King, 2018). While ethnodevelopment is a well-established concept in the development studies literature, it is not so well known in the field of tourism studies (King & Borges de Lima, 2018). Ethnodevelopment is a type of development that recognises diversity as a fundamental resource for addressing the endogenous needs of specific ethnic groups and minority communities and as such, is concerned with the maintenance of ethnic diversity (Stavenhagen, 1986). It recognises the importance of the cultural dimension in identity formation, the need for the empowerment of marginalised groups, and the importance of encouraging local involvement in decision-making. It focuses on the crucial roles that culture and Indigenous or traditional knowledge and skills play in the local acceptance and success of development projects and therefore, the sustainability of local livelihoods (King & Borges de Lima, 2018). In compliance with the concepts of Indigenous community and communality and with the absolute definition of equity, ethnodevelopment favours reciprocity, collective participation, common labour and the well-being of the community above that of the individual; thus, cultural processes are determined by and managed for the social group. In ethnodevelopment, the preservation and protection of local identity and tradition occurs within the group's own structures and institutions, as well as within a local context that creates linkages to the global (Warnholtz & Barkin, 2018). For Indigenous communities, this form of development potentially offers a way to define Indigenous development as community-focused, culturally-driven and rooted in ethnicity and diversity.

Indigenous knowledge in governance, development and research

As there is a shift towards developing, analysing and understanding Indigenous tourism from more Indigenous-centric models, such as through ethnodevelopment, there is also a shift towards greater use of Indigenous knowledge in Indigenous tourism and its development. Given that local communities are the custodians of cultural heritage resources and Indigenous knowledge systems, it is cogent that these communities should control the way their cultural heritage is passed onto others through mechanisms such as tourism, and at the same time be active participants in the process (Nkwanyana, 2018). Indeed, scholars and practitioners are increasingly recognising that in the development of Indigenous tourism initiatives, the employment of top-down strategies focused predominantly towards the provision of tourist facilities lead to the sustained perpetuation of stereotyped ideas of Indigenous peoples (Espeso-Molinero et al., 2016; Whitford & Ruhanen, 2010). Such strategies also silence those Indigenous voices on the margins, thus potentially perpetuating the perception of Indigenous knowledge as inferior, static, simple and primitive (Buzinde, Vandever, & Nyaupane, 2017). When strategies imposed from above disregard local cultural identities, local knowledge systems, marginal voices and specific realities of Indigenous entrepreneurs, the chances of success are reduced considerably (Whitford & Ruhanen, 2010). Thus, in developing Indigenous enterprises and strategies, the full participation and involvement of Indigenous peoples through collaborative strategies that foster knowledge dialogue between scientific and traditional wisdom offer a viable solution to the current implementation gaps (Espeso-Molinero et al., 2016).

With burgeoning studies showcasing the complexity and sophistication of many Indigenous natural resource management systems, researchers are also becoming more cognisant that Indigenous knowledge augurs well for sustainable development studies and environmental conservation (Borges de Lima & King, 2018; Buzinde et al., 2017; Espeso-Molinero et al., 2016; Pereiro, 2016; Scherrer & Doohan, 2013). Indeed, it is becoming increasingly common to not only have Indigenous people employed by national conservation agencies and involved in environmental stewardship, but also to have protected-areas' boards co-managed by, or composed exclusively of, Indigenous membership (Maraud & Guyot, 2016). The inclusion of Indigenous knowledge in the packaging of tourism products also increases the 'authenticity' of those products while having the potential of enriching the experiences of tourists (Nkwanyana, 2018). From a cultural sustainability perspective, the inclusion of Indigenous knowledge in tourism development and participation in cultural activities by Indigenous peoples are also important ways of passing cultural heritage from one generation to the next (Nkwanyana, 2018).

Recognition of this 'strong relationship between successful sustainable development projects and Indigenous knowledge' (Buzinde et al., 2017, p. 45) suggests that, in future research and development endeavours, the dividing practice of separating Indigenous and scientific knowledge should be avoided (Dyll, 2018). As recognised by Dyll (2018), the rethinking of indigeneity requires removing imperialist epistemological assumptions of researchers as 'experts' and of Indigenous knowledge as a barrier to progress. Indeed, tourism needs to become a space that is seen as one that provides the opportunity for potential mutual learning (Scherrer & Doohan, 2013). In order to overcome the historical marginalisation of Indigenous peoples, there is a need for more collaborative research between academics and Indigenous communities, as well as strategies that foster the exchange of ideas between traditional and scientific knowledge (Espeso-Molinero et al., 2016). Research processes in future development initiatives must therefore be undertaken with multiple partners and the academic methodologies used must enable cultural factors to be considered and Indigenous knowledge as a significant element of the theoretical scaffolding of the study.

Recognition of the positive correlation between successful sustainable development projects and Indigenous knowledge also provides support for community-based approaches to tourism. Such approaches regard Indigenous knowledge as the basis for grassroots decision-making processes and therefore involve local communities and groups in problem identification and solution generation (Buzinde et al., 2017). Pookhao et al. (2018) suggest that bottom-up, community-based approaches to tourism development and governance are also viable means to improve the quality of life for Indigenous people through local empowerment.

While community-based tourism approaches have been utilised extensively within tourism contexts (Buzinde et al., 2017) and are suggested as a way to better include Indigenous voices in tourism, when such approaches are implemented as elements of national development policies they have the tendency to fail (Warnholtz & Barkin, 2018). Warnholtz and Barkin (2018) suggest that the issues in determining the success or failure of Indigenous community-based tourism projects are also much more complex than have hitherto been realised in the context of the generally accepted tourism development paradigm. The authors identify a lack of research comparing Indigenous community-based tourism interventions from the perspectives of the host community on the one hand, and those of the development policy-formulators and practitioners on the other hand. They suggest that in designing successful interventions on behalf of a host community the focus of policy-thinking and planning should incorporate spaces for local empowerment and self-determination in order to meet the needs and expectations of that community, along with those of the tourists. They also suggest that projects need to be based more firmly on the concept of local community-based development, as well as on a more inclusive definition of tourism as a multi-faceted and multi-dimensional phenomenon. Pookhao et al. (2018) add that studies into Indigenous community-based tourism enterprises and initiatives need to also take a place-based ethnographic focus that is linked to the broader context of supranational institutions, governmental organisations, NGOs, tourists and local entrepreneurs because all of these greatly influence the success or otherwise of the tourism development projects. Indeed, initiatives that maintain dialogue both within and beyond the local communities and which are established on the basis of community-based, collaborative and participatory decision-making have been observed to more successfully encourage cultural revival, the positive evaluation of local culture and identity, the inclusion of local communities in the development of their own futures, and their involvement in environmental stewardship (Borges de Lima & King, 2018).

Special issue papers

The following seven papers in the issue explore Indigenous cultural heritage and tourism in different countries and contexts with contributions from Latin America, North America, Scandinavia, Africa and Australia.

The first paper, 'Evolution of indigenous tourism among the Lacandon of Chiapas: an application of Weaver's model', by Pilar Espeso-Molinero and María José Pastor-Alfonso uses Weaver's six-stage Indigenous tourism model (pre-European in situ control; in situ exposure; ex situ exhibitionism and exploitation; in situ exhibitionism and exploitation; in situ quasi-empowerment; and ex situ quasi-empowerment) to assess the development of tourism in the Lacandon Maya (Hach Winik) region of Chiapas, Mexico. While the model has applicability for the Lacandon context, key differences due to the Latin American context were observed.

Tatjana Thimm's paper, 'Cultural sustainability – a framework for Aboriginal tourism in British Columbia', uses a range of primary and secondary data sources to develop a framework to assess the cultural sustainability of Aboriginal tourism in British Columbia. The framework highlights the importance of cultural sustainability for achieving success through tourism while also acknowledging the protection of human rights, good self-governance, identity, control of land, authentic and market-ready tourism products.

An African perspective on cultural heritage tourism is offered in Kokel Melubo and Anna Carr's paper, 'Developing indigenous tourism in the bomas: Critiquing issues from within the Maasai community in Tanzania'. Their study draws on the insights of local stakeholders including the Maasai community to identify the critical challenges for providing indigenous cultural tourism in the Ngorongoro Conservation Area of Tanzania. As was the case in Thimm's Canadian context, the study findings here highlight a number of challenges in developing successful Indigenous cultural tourism and Melubo and Carr discuss the importance of targeted community based strategies to facilitate the development of Maasai cultural tourism.

Bryan Grimwood, Meghan Muldoon and Zachary Stevens' paper on 'Settler colonialism, Indigenous cultures, and the promotional landscape of tourism in Ontario, Canada's "Near North"' discusses the notion of settler colonialism that have displaced Indigenous Peoples and cultural connections to land. Against this backdrop, they examine how Indigenous culture is portrayed in the promotion of northern Ontario; home for generations of the Anishinaabeg and Haudenosaunee First Nations. They use a critical discourse analysis of websites and case study sites to explore the role of tourism in maintaining and revitalising Indigenous culture. Importantly the paper contributes to critiques of settler colonial power relations and the impact on tourism and Indigenous cultures.

The fifth paper in the special issue, 'Authenticity as a compromise: A critical discourse analysis of Sami tourism websites' by Cecilla de Bernardi also explores the imagery associated with Indigenous cultural tourism and the prevalence of colonial images which no longer align with contemporary Indigenous culture. This study examines the connection between tourism marketing, authenticity and representation of cultural heritage through an examination of the websites of Sámi tourism companies from Sweden. The paper highlights aspects of the different discourses on representations and authenticity via the main channel to reach the tourist market.

From Australia, Trinidad Espinosa's paper, 'Indigenous tourism in Australia: understanding the link between cultural heritage and intention to participate using the means-end chain theory' uses means-end chain theory and mixed methods based on photographs to examine the link between visitor perceptions about Australian Indigenous tourism attributes and their intention to participate. Visitor perceptions of two Australian Indigenous tourism activities, focused on traditional and contemporary culture respectively, were examined. As with de Bernardi's study, it was found that some visitors associate Indigenous tourism with 'traditional' imagery but increasingly, participants focused on the contemporary aspects of Indigenous culture. Espinosa discusses the importance of Indigenous tourism stakeholders monitor and protecting their historical and evolving cultural heritage, particularly intangible heritage.

The final paper in the special issue by Chance Finegan provides a slightly different perspective to the previous papers. In 'The interpreter as researcher: Ethical heritage interpretation in Indigenous contexts', Finegan integrates Indigenous research and heritage interpretation scholarship to discuss how interpretation should be underpinned by research that is congruent with Indigenous expectations for research ethics and methods. The paper focuses on Indigenous research methods and ethics, and discusses the connections between Western knowledge and colonialism, to highlight the issues for interpreters working with Indigenous content.

Conclusion

Travellers have long sought to experience the unfamiliar and exotic cultures and traditions of Indigenous peoples, with early examples of Indigenous tourism in the United States, Canada, Scandinavia, Australia, New Zealand and countries throughout Asia and Latin America. Similarly, contemporary travellers demonstrate a desire to seek out opportunities to experience Indigenous peoples and their culture (Nielsen & Wilson, 2012). Thus, we are witnessing world-wide growth in the awareness of, and interest in, Indigenous cultures, traditions, histories and knowledges.

Tourism has also long been promoted for its potential to revitalise Indigenous communities (Dyer, Aberdeen, & Schuler, 2003) and to sustain Indigenous cultural heritage. However, the nexus between culture and tourism is not always a positive one. Indigenous cultures have been the subject of academic writers and philosophers (predominantly anthropologists) for hundreds of years and for nearly half a century from a tourism lens. This paper has provided an overview of the main issues being addressed in contemporary research that explores the nexus between cultural heritage and Indigenous tourism and introduced the papers that form this special issue of the Journal of Heritage Tourism. This review, together with the special issue papers, have shown the myriad of ways that tourism can affect Indigenous culture and cultural identity. While modernisation and globalisation has led to the erosion of many Indigenous cultures and cultural heritage, tourism is arguably one mechanism through which to revive these cultures. However, of ongoing concern, is the paradoxical condition placed upon many Indigenous communities involved in tourism. This paradox sees Indigenous people negotiating the desire to develop socio-economically, while at the same time needing to 'stand still in time' to meet the needs of tourists. This conundrum has been contemplated by many researchers, and while a growing number of researchers are calling for the conceptualisation of 'authentic' cultural practices as constantly 'in transition', tourists still evaluate their Indigenous tourism experiences against expectations shaped by popular stereotypical representations of archaic practices. Indeed, around the world, many Indigenous people still find themselves with very little control of the imagery that commodifies their indigeneity and which stereotypes them as timelessly authentic. Here, therefore, lies a challenge for researchers: how can such entrenched stereotypes be challenged? And how can this be achieved on a global scale? Too often questions of cultural ownership and the 'right' to share culture still neglects the very people whose culture is being shared. Indeed there is a need to listen more carefully to minority voices. There is also a need to find better ways for giving these voices more prominence so they are better able to challenge dominant narratives, and as researchers, we need to be more cognisant of not only listening to Indigenous voices, but partnering with and co-creating our research to ensure Western analyses of tourist systems is not the domineering paradigm.

From a sustainability perspective it is also important that we as researchers consider alternative ways to development and sustainability. Such alternatives need to take a much greater community-driven approach and much less of a Western-centric view of what constitutes sustainable development, thus shifting focus from the immediate needs of visitors and the tourism industry in general to instead consider the cultural values at stake within tourism development. Questions thus arise around how to more meaningfully engage with Indigenous peoples' relationship with the land and by extension, the cultural space in which tourism initiatives operate, as meaningful engagement is arguably sustainable engagement.

With the increased adoption of ethnodevelopment, we are currently witnessing a greater shift towards developing, analysing and understanding Indigenous tourism from more Indigenous-centric models. There is also a shift towards greater use of Indigenous knowledge in Indigenous tourism and its development. While such a shift is positive and needed to ensure the continued integration of emic perspectives of Indigenous peoples, use of Indigenous-centric research and knowledge models and application of Indigenous knowledge must become the norm in research and development that focuses on Indigenous cultural heritage in tourism and Indigenous development. As suggested by Dyll (2018, p. 333), 'the application of critical and Indigenous qualitative methodologies necessitates the self-reflexive problematisation of Self–Other and researcher–researched relationships. Its objective is to make a case for research participants (normally known as 'informants', 'subjects', 'sources') to be included in certain kinds of studies as co-authors and co-researchers.' As researchers, when we are involved in studies examining Indigenous issues in tourism and cultural heritage, we must all adapt our research practices to ensure they are ethical, transformative, participatory, and committed to dialogue with the Indigenous people who are partners in the research. Indeed, to overcome the historical marginalisation of Indigenous peoples, we need to make greater effort towards the collaborative co-creation of research between academics and Indigenous communities, while also adopting strategies that foster the exchange of ideas between traditional and scientific knowledge.

Disclosure statement

No potential conflict of interest was reported by the authors.

References

Altman, J. C., & Finlayson, J. (1992). *Aborigines, tourism and sustainable development*. Canberra: Centre for Aboriginal Economic Policy Research, Australian National University. Retrieved from https://www.jcu.edu.au/__data/assets/pdf_file/0011/122204/jcudev_012859.pdf

Borges de Lima, I., & King, V. T. (Eds.). (2018). *Tourism and ethnodevelopment: Inclusion, empowerment and self-determination*. New York, NY: Routledge.

Bott, E. (2018). Among the piranhas: The troubling lifespan of ethnic tropes in "tribal" tourism to Vietnam. *Journal of Sustainable Tourism*, *26*(8), 1291–1307.

Bresner, K. (2014). Sharing identity through indigenous tourism: Osoyoos Indian Band's Nk'Mip Desert Cultural Centre. *Anthropologica*, *56*(1), 135–150.

Bunten, A. (2008). Sharing culture or selling out? Developing the commodified persona in the heritage industry. *The American Ethnologist*, *35*(3), 380–395.

Bunten, A. (2014). Deriding demand: Indigenous imaginaries in tourism. In N. B. Salazar & H. H. Graburn (Eds.), *Tourism imaginaries: Anthropological approaches* (pp. 80–102). New York, NY: Berghahn Books.

Burns, P. M. (2006). Social identities and the cultural politics of tourism. In P. M. Burns & M. Novelli (Eds.), *Tourism and social identities: Global frameworks and local realities* (pp. 13–24). Amsterdam: Elsevier.

Butler, S. (2017). 'Temporary belonging': Indigenous cultural tourism and community art centres. In S. Carson, & M. Pennings (Eds.), *Performing cultural tourism. Communities, tourists and creative practices* (pp. 27–42). New York, NY: Routledge.

Buzinde, C. N., Vandever, V., & Nyaupane, G. (2017). Native American communities and community development: The case of Navajo Nation. In S. Carson, & M. Pennings (Eds.), *Performing cultural tourism: Communities, tourists and creative practices* (pp. 44–60). New York, NY: Routledge.

Campos, M. V., Scott, N., & Breaky, N. (2014). Ecotouris: A new challenge for protected rainforest areas in Chile. In B. Prideaux (Ed.), *Rainforest tourism, conservation and management: Challenges for sustainable development* (pp. 134–145). New York, NY: Earthscan/Routledge.

Carr, A., Ruhanen, L., & Whitford, M. (2016). Indigenous peoples and tourism: The challenges and opportunities for sustainable tourism. *Journal of Sustainable Tourism*, *24*(8–9), 1067–1079.

Cohen, E. (1988). Authenticity and commoditization in tourism. *Annals of Tourism Research*, *15*(3), 371–386.

Coronado, G. (2014). Selling culture? Between commoditisation and cultural control in Indigenous alternative tourism. *Pasos: Revista de Turismo y Patrimonio Cultural*, *12*(1), 11–28.

de Kadt, E. (1979a). Social planning for tourism in the developing countries. *Annals of Tourism Research*, *6*(1), 36–48.

De Kadt, E. (1979b). *Tourism: Passport to development? Perspectives on the social and cultural effects of tourism in developing countries*. New York, NY: Oxford Universty Press.

Dyer, P., Aberdeen, L., & Schuler, S. (2003). Tourism impacts on an Australian indigenous community: A Djabugay case study. *Tourism Management*, *24*(1), 83–95.

Dyll, L. (2018). Indigenous environmental knowledge and challenging dualisms in development: Observations from the Kalahari. *Development in Practice*, *28*(3), 332–344.

Einar Johansen, T., & Mehmetoglu, M. (2011). Indigenous tourism from a visitor's perspective: An empirical examination of Valene L. Smith's 4Hs at a sámi festival in Norway. *Journal of Heritage Tourism*, *6*(2), 129–141.

Espeso-Molinero, P., Carlisle, S., & Pastor-Alfonso, M. J. (2016). Knowledge dialogue through Indigenous tourism product design: A collaborative research process with the Lacandon of Chiapas, Mexico. *Journal of Sustainable Tourism*, *24*(8–9), 1331–1349.

Espinosa Abascal, T., Fluker, M., & Jiang, M. (2015). Domestic demand for Indigenous tourism in Australia: Understanding motivations, barriers, and implications for future development. *Journal of Heritage Tourism*, *10* (1), 1–20.

Giblin, J. D. (2017). Touring and obscuring poverty: Urban and rural cultural-heritage tourism. *Heritage & Society*, *10* (2), 128–146.

Heldt Cassel, S., & Maureira, T. M. (2017). Performing identity and culture in Indigenous tourism – A study of Indigenous communities in Québec, Canada. *Journal of Tourism and Cultural Change*, *15*(1), 1–14.

Hinch, T., & Butler, R. (2007). Introduction: Revisiting common ground. In R. Butler & T. Hinch (Eds.), *Tourism and indigenous peoples* (pp. 1–12). Oxford: Butterworth-Heinemann.

Hsu, P.-H. (2018). Tourist perceptions of living authenticity in indigenous tourism destinations: The case of Smangus Village in Taiwan. *Tourism, Culture & Communication*, *18*(2), 117–131.

Hsu, P.-H., & Nilep, C. (2018). Authenticity in indigenous tourism: The provider's perspective. *International Journal of Critical Indigenous Studies*, *8*(2), 16–28.

Kagumba, A. K. (2013). *Indigenous cultural tourism and the discourse of development among the Batwa of Mgahinga, South Western Uganda* (Master of Philosophy in Indigenous Studies). University of Tromsø, Norway.

Kessy, D. (2017). Involving indigenous community in tourism activities: Strategies and interventions. *African Journal of Hospitality, Tourism and Leisure*, *6*(4), 1–9.

King, D. A., & Stewart, W. P. (1996). Ecotourism and commodification: Protecting people and places. *Biodiversity and Conservation*, *5*(3), 293–305.

King, V. T., & Borges de Lima, I. (2018). Tourism and ethnodevelopment: An introduction. In I. Borges de Lima & V. T. King (Eds.), *Tourism and ethnodevelopment: Inclusion, empowerment and self-determination* (pp. 3–24). New York, NY: Routledge.

Korstanje, M. (2012). Reconsidering cultural tourism: An anthropologist's perspective. *Journal of Heritage Tourism*, *7* (2), 179–184.

Liu, Z. (2018). Searching for a lost aura: A Naxi Dongba's spatial practices and space remaking in touristic commoditization. *Journal of Tourism and Cultural Change*, *16*(4), 348–364.

Mackley-Crump, J. (2016). From Private Performance to the public stage: Reconsidering 'staged authenticity' and 'traditional' performances at the Pasifika festival. *Anthropological Forum*, *26*(2), 155–176.

Maraud, S., & Guyot, S. (2016). Mobilization of imaginaries to build Nordic Indigenous natures. *Polar Geography*, *39* (3), 196–216.

McIntosh, A. J., Hinch, T., & Ingram, T. (2002). Cultural identity and tourism. *International Journal of Arts Management*, *4*(2), 39–42.

Nielsen, N., & Wilson, E. (2012). From invisible to indigenous-driven: A critical typology of research in indigenous tourism. *Journal of Hospitality and Tourism Management*, *19*(1), 67–75.

Nkwanyana, S. M. (2018). Conserving cultural heritage and the usage of indigenous knowledge: A case study of the Zululand District Municipality, Kwazulu-Natal. *African Journal of Hospitality, Tourism and Leisure*, *7*(2), 1–12.

Ona, J. G., & Solis, L. S. L. (2017). Challenges and opportunities for the development and promotion of Ibaloy crafts as tourism products. *International Journal of Culture, Tourism and Hospitality Research*, *11*(4), 593–607.

Pereiro, X. (2016). A review of Indigenous tourism in Latin America: Reflections on an anthropological study of Guna tourism (Panama). *Journal of Sustainable Tourism*, *24*(8-9), 1121–1138.

Pookhao, N., Bushell, R., Hawkins, M., & Staiff, R. (2018). Empowerment through community-based ecotourism in a globalised world global-local nexus – Three Thai villages as case studies. In I. Borges de Lima & V. T. King (Eds.), *Tourism and ethnodevelopment: Inclusion, empowerment and self-determination* (pp. 65–81). New York, NY: Routledge.

Ranasinghe, R., & Cheng, L. (2018). Tourism-induced mobilities and transformation of indigenous cultures: Where is the Vedda community in Sri Lanka heading to? *Journal of Tourism and Cultural Change, 16*(5), 521–538.

Richards, G. (2018). Cultural tourism: A review of recent research and trends. *Journal of Hospitality and Tourism Management, 36*, 12–21.

Rights and Resources Initiative. (2015). *Who owns the world's land? A global baseline of formally recognized indigenous and community land rights*. Washing, DC: Author. Retrieved from http://rightsandresources.org/wp-content/uploads/GlobalBaseline_complete_web.pdf

Ruhanen, L., & Whitford, M. (2018). Racism as an inhibitor to the organisational legitimacy of Indigenous tourism businesses in Australia. *Current Issues in Tourism, 21*(15), 1728–1742.

Ruhanen, L., Whitford, M., & McLennan, C. L. (2015). Indigenous tourism in Australia: Time for a reality check. *Tourism Management, 48*, 73–83.

Ryan, C., & Aiken, M. (Eds.). (2015). *Indigenous tourism: The commodification and management of culture*. Oxford: Elsevier

Saarinen, J. (2013). Indigenous tourism and the challenge of sustainability. In M. Smith & G. Richards (Eds.), *The Routledge handbook of cultural tourism* (pp. 220–226). New York, NY: Routledge.

Saugestad, S. (2000). Dilemmas in Norwegian development assistance to indigenous peoples. *Forum for Development Studies, 27*(2), 205–234.

Saugestad, S. (2008). Beyond the 'Columbus context': New challenges as the indigenous discourse is applied to Africa. In H. Minde (Ed.), *Indigenous peoples: Self-determination, knowledge, indigeneity* (pp. 157–173). Delft: Eburon Academic Publishers.

Scherrer, P., & Doohan, K. (2013). It's not about believing': Exploring the transformative potential of cultural acknowledgement in an Indigenous tourism context. *Asia Pacific Viewpoint, 54*(2), 158–170.

Stavenhagen, R. (1986). Ethnodevelopment: A neglected dimension in development thinking. In R. Apthorpe, & A. Krahl (Eds.), *Development studies: Critique and renewal* (pp. 71–94). Leiden: E.J. Brill.

Stronza, A. (2008). Through a new mirror: Reflections on tourism and identity in the Amazon. *Human Organization, 67*(3), 244–257.

The World Bank. (2018). Indigenous peoples. Retrieved from https://www.worldbank.org/en/topic/indigenouspeoples

Thompson-Carr, A. (2013). Maori tourism: A case study of managing indigenous cultural values. In M. Smith & G. Richards (Eds.), *The Routledge handbook of cultural tourism* (pp. 227–235). New York, NY: Routledge.

Tomaselli, K. G. (Ed.). (2012). *Cultural tourism and identity: Rethinking indigeneity*. Leiden: Martinus Nijhoff Publishers.

Travesi, C. (2018). Knowing and being known: Approaching Australian indigenous tourism through aboriginal and non-aboriginal politics of knowing. *Anthropological Forum, 28*(3), 275–292.

United Nations. (2017). *State of the world's indigenous peoples: Education*. 3rd Volume. New York, NY: Author. Retrieved from https://www.un.org/development/desa/indigenouspeoples/wp-content/uploads/sites/19/2017/12/State-of-Worlds-Indigenous-Peoples_III_WEB2018.pdf

Warnholtz, G., & Barkin, D. (2017). Development for whom? Tourism used as a social intervention for the development of indigenous/rural communities in natural protected areas. In I. Borges de Lima & V. T. King (Eds.), *Tourism and ethnodevelopment: Inclusion, empowerment and self-determination* (pp. 49–65). New York, NY: Routledge.

Warnholtz, G., & Barkin, D. (2018). Development for whom? Tourism used as a social intervention for the development of indigenous/rural communities in natural protected areas. In I. Borges de Lima & V. T. King (Eds.), *Tourism and ethnodevelopment: Inclusion, empowerment and self-determination* (pp. 27–43). New York, NY: Routledge.

Weaver, D. B. (2016). Beyond convention: Reimagining indigenous tourism. *Tourism Geographies, 18*(1), 28–37.

Whitford, M., & Ruhanen, L. (2010). Australian Indigenous tourism policy: Practical and sustainable policies? *Journal of Sustainable Tourism, 18*(4), 475–496.

Whitford, M., & Ruhanen, L. (2016). Indigenous tourism research, past and present: Where to from here? *Journal of Sustainable Tourism, 24*(8–9), 1080–1099.

Whitney-Squire, K. (2016). Sustaining local language relationships through indigenous community-based tourism initiatives. *Journal of Sustainable Tourism, 24*(8–9), 1156–1176.

Williams, L. K., & Gonzalez, V. V. (2017). Indigeneity, sovereignty, sustainability and cultural tourism: Hosts and hostages at 'Iolani Palace, Hawai'i. *Journal of Sustainable Tourism, 25*(5), 668–683.

Yang, L. (2012). Tourists' perceptions of ethnic tourism in Lugu Lake, Yunnan, China. *Journal of Heritage Tourism, 7*(1), 59–81.

Zeppel, H. (2010). Managing cultural values in sustainable tourism: Conflicts in protected areas. *Tourism and Hospitality Research, 10*(2), 93–104.

Evolution of indigenous tourism among the Lacandon of Chiapas: an application of Weaver's model

Pilar Espeso-Molinero and María José Pastor-Alfonso

ABSTRACT

In this paper, Weaver's six-stage indigenous tourism model is applied to the Lacandon Maya (Hach Winik) of Chiapas, Mexico. Based on a comprehensive review of the anthropological and historical literature on this indigenous group, combined with longitudinal ethnographic and collaborative research performed with tourism entrepreneurs, the Lacandon tourism experience is assessed from the pre-European period until present. By analysing a case study of indigenous tourism in Mexico, a developing country in another geographical region and with a different colonial past, this work supplements Weavers' perspective. The results show that the fourth and fifth stages of Weaver's model coincide in this case study, while the sixth stage is still incomplete. Although the Lacandon case has its peculiarities and bearing in mind that several different factors should be considered in the Latin American context, the model proves to be an interesting tool for indigenous tourism analysis in developing countries.

Introduction

Since 1974, when Valene Smith organised the first symposium on tourism and anthropology, the indigenous tourism concept has received greater attention from academics and practitioners (Butler & Hinch, 1996; Buultjens & Fuller, 2007; Carr, Ruhanen, & Whitford, 2016; Lemelin & Blangy, 2009; Ryan & Aicken, 2005; Smith, 1977). The growing body of literature studies the phenomenon from the perspectives of hosts (McIntosh, Zygadlo, & Matunga, 2004) and visitors (McIntosh & Ryan, 2007) alike. And although very few works have attempted to conceptualise the evolution of the relationship between tourism and indigenous peoples, a notable exception is Weaver's (2010) paper on indigenous tourism stages and their implications for sustainability.

After studying more than 200 academic and professional documents on indigenous tourism, Weaver developed a six-stage evolution model of the relationship between tourism and indigenous peoples: (1) pre-European in situ control; (2) in situ exposure; (3) ex situ exhibitionism and exploitation; (4) in situ exhibitionism and exploitation; (5) in situ quasi-empowerment; and (6) ex situ quasi-empowerment. In his analysis of the dynamics of empowerment and its implications for sustainability, he acknowledges two main limitations: the absence of indigenous 'voices' among his sources and a corpus containing only English-language documents, thus restricting the focus of his research to cases in Australia, Canada, New Zealand and the United States. He also considers that it 'is unlikely that case studies from less economically developed regions in Latin America, Africa, Asia and the South Pacific in particular will reveal the full six-stage progression' (Weaver, 2010, p. 56).

Therefore, an attempt is made here to overcome those limitations by applying Weaver's model to a case study of the evolution of tourism among the Lacandon Maya Indians (also known as the Hach Winik, i.e. 'true men') living in the jungles of Chiapas, near Mexico's southern border with Guatemala.

Following Weaver's example, this paper examines both written ethnographies – the Lacandon Maya have received considerable scholarly attention and consequently there is an extensive body of anthropological and historical literature that facilitates the understanding of the evolution of tourism in their communities – and government documents. On the other hand, thanks to the long-term collaboration between the Intercultural University of Chiapas (UNICH) in Mexico, the University of Alicante (UA) in Spain and the tourism entrepreneurs of the Lacandon Jungle, a collaborative team of researchers have produced accounts of different aspects of the relationship between the Lacandon communities and the tourism industry, thus providing primary data for this analysis.

The economy of the Lacandon Maya was based on subsistence farming, hunting and fishing until tourism took centre stage. As will be seen, its evolution in the area has much in common with Weaver's stages. However, the political and social idiosyncrasies of this small but powerful group have created different resistance dynamics that deserve further discussion in order to add a new perspective to Weaver's model.

The Lacandon people

Despite being one of the least numerous indigenous groups in Mesoamerica, the Lacandon Maya have constantly fascinated visitors and scholars alike. They live in one of the planet's most biodiverse areas and, over time, have acquired a fund of knowledge in agriculture, botany, natural medicine, etc. Their customs and ancestral beliefs continue to arose the interest of academics, the attention of the media and, lately, also the curiosity of tourists, eager to discover a reality very different from their own.

Enhancing the destination's appeal, the Lacandon Jungle encompasses the legacy of the Classic Maya, who left behind remains of important cities such as Palenque, Bonampak and Yaxchilán. Since the first expeditions reached the area at the end of the nineteenth century, the Lacandon Maya have offered their hospitality to both friends and strangers. More recently, with the active support of the state and national governments, they have begun to convert their traditional hospitality into a tourist business.

Habitat and history

According to the French ethnologist Soustelle's description in 1935 (1935/2008), 'the entire Lacandon territory is muddy, swampy, extremely humid' (p. 15) and is bordered by a mighty river containing the waters of numerous tributaries that run through the jungle. Multiple lakes and lagoons also dot the territory, 'like mirrors of jade in the middle of the forests' (Soustelle, 1935/2008, p. 15), making it one of the areas with the highest concentration of surface water in Mexico (Pastor-Alfonso, 2012).

The Hach Winik have been living in the jungles of Chiapas for the past three centuries. Their dominions, language and traditions, with obvious Mayan connections, led the first explores and scholars to assume that they were the Lacam-tun people, to wit, direct descendants of the Maya. However, historical (de Vos, 1980; Palka, 2005) and linguistic research (Bruce, 1968) has finally dispelled the myth of historical continuity with the pre-Columbian past in Mexico, demonstrating that they are two different groups. The Lacam-tun people, called the 'Lacandon' by the Spanish colonisers, were a bellicose pre-Hispanic indigenous group that fought relentlessly for their freedom against the 'usurpers'[1] until the point of total extinction (de Vos, 1980). After subduing the natives, the Spaniards encountered difficulties in colonising the area and soon left. The abandoned settlements were then occupied by other Mayan groups that had been pushed out of other areas of the ancient Mayan empire by colonisation (Palka, 2005). Little else is known about their origins.

The following centuries passed in a very similar manner for the Lacandon. Their traditional settlements were scattered throughout the rainforest in two distinctive groups, north and south, with some linguistic and cultural differences, until the Mexican land reforms in the 1950s opened up the jungle to colonisation and industrial exploitation. Encouraged by agrarian reforms, Tzeltal, Tzotzil and Chol immigrants from the highlands and the northern reaches of Chiapas begin to settle in the lowlands, bringing with them their own traditional ways of subsistence unrelated to the jungle economy.

Migratory pressure combined with political incentives forced the Lacandon to unify their settlements. Consequently, the dispersed pattern of family clusters disappeared to be replaced by the modern-day communities of Nahá, Metzabok (north) and Lacanjá Chansayab (south).

Development and tourism among the Lacandon people

Although logging and cattle ranching have transformed the traditional way of life of the Lacandon, the most obvious changes have been brought about by the development of tourism in the area. Government support for tourism as a socio-economic alternative and a way of protecting the environment began at the end of the twentieth century. The close collaboration that the Lacandon have maintained with the state and national authorities since the 1970s has placed this small group in a privileged and unprecedented position as regards other indigenous groups. Despite their small population, they have received generous funding from the different administrations (Pastor-Alfonso, Gómez López, & Espeso-Molinero, 2012). Furthermore, they have been granted privileges, such as archaeological returns (access to the Bonampak Natural Monument), control over three protected areas (Metzabok, Nahá and Montes Azules), and extensive lands with diverse natural and ecotourism attractions (waterfalls, lagoons, etc.). Through different public programmes, in conjunction with local initiatives and with the support of NGOs and universities, they have consolidated a wide range of nature tourism facilities and services, which for many families have replaced traditional economic activities (Pastor-Alfonso, 2012).

The Lacandon in the academic literature

Following the arrival of the ethnologist Tozzer (1907) at the beginning of the twentieth century, there have been expeditions to the Lacandon Jungle without pause. In addition to numerous general monographs on the Hach Winik (Boremanse, 1998; Eroza Solana, 2006; Marion Singer, 1999; McGee, 2002; Perera & Bruce, 1985), anthropologists, linguists, ethnologists and historians have focused their attention on specific topics, such as the Mayan language of the Lacandon (Bruce, 1968), their cosmology (Marion Singer, 2000), religious rituals (Boremanse, 2007; McGee, 1990; Tozzer, 1907) and funerals rites (Blom, 1954). Researchers have also concentrated on their impressive oral tradition (Bruce, 1976), with its rich mythology (Boremanse, 1989; Bruce, 1979), their social, family and gender organisation (Marion Singer, 1999; McGee & González, 1999), their ethnobotanical knowledge and use of medicinal plants (Kashanipour & McGee, 2004), their subsistence economy based on agriculture and hunting (Baer & Merrifield, 1972; Diemont & Martin, 2009) and the technologies that they have developed in this respect (Marion Singer, 1991; Nations & Clark, 1983).

Similarly, many works address the controversy surrounding the historical evolution of the Lacandon Maya (Baer & Merrifield, 1972; de Vos, 1980, 2002; Palka, 2005). And in recent years there has been a clear trend towards studies focusing on traditional ecological knowledge (Diemont & Martin, 2009), the political controversy over land rights (Calleros-Rodríguez, 2014; Trench, 2008), environmental management (Durand, Figueroa, & Trench, 2012; Reygadas, Ramos, Montoya Gómez, Hernández, & Velasco, 2006) and the arrival of tourism in the Lacandon communities (Espeso-Molinero, Carlisle, & Pastor-Alfonso, 2016; Espeso-Molinero, 2018; Hernandez Cruz, Bello-Baltazar, Montoya Gómez, & Estrada Lugo, 2005; López & Palomino, 2008; Pastor-Alfonso & Gómez López, 2010; Pastor-Alfonso, 2012; Valle-García, 2014).

In this extensive and varied production, there is only one issue that crops up time and again: the advance of global acculturation and the inexorable loss of Lacandon mores. Evangelicalism and formal education, with its nationalist symbols and biased views on the historical past, have gradually encroached on the mythical Lacandon cosmology. As many authors (Boremanse, 2006; McGee, 2002; Valle-García, 2014) have already indicated, agricultural transformation as regards techniques and crops and diet changes are gradually eroding the indigenous people's gastronomic, botanical and ethnomedicinal knowledge base, leading to irreversible cultural changes. In contrast, only a few studies have underscored the opportunities afforded by tourism-related activities for cultural revival (Espeso-Molinero et al., 2016; Espeso-Molinero, 2018).

Lacandon tourism according to Weaver's model

In 2010, Weaver proposed a model to 'assess the evolution of the relationship between tourism and indigenous peoples' (p. 43). Employing the control/theme matrix developed by Hinch and Butler in 1996 as a 'framework for contextualizing and visualizing' (p. 46) the basic variables, his model represents an ideal six-stage evolution.

Stage one: in situ control

The first stage of Weaver's model considers the phase prior to European colonisation and would refer to activities carried out by indigenous peoples as part of their ancestral customs. Ceremonial reunions and gift exchanges between neighbouring groups or *corroborees*, celebrated as much between the North American Indians (*potlatch*) as between the Maori (*koha*), involved displacements and situations that today would be recognised as tourism activities. This 'pre-European' stage was controlled in situ by indigenous groups.

In the Lacandon context, the ethnocide of the Lacam-tun group prevents us from understanding indigenous meeting, ceremonial reunion or gift exchange practices prior to the Spanish colonisation. Among the Hach Winik, there are no ceremonial meetings among neighbours. For many years, this indigenous group was accustomed to war and perhaps because of this, not especially inclined to interact with other groups (Boremanse, 1998). Their constant need to defend themselves against the Spanish military, the Catholic missionaries, other indigenous groups and even other Lacandon communities (the Lacandon of the south were renowned for both their violent nature and habitual practice of abducting women for marital purposes) prompted especially those from the north to disperse and seek isolation. So, in principle, it seems that those from the north preferred to avoid contact with the inhabitants of other settlements and visits were only paid for specific reasons (e.g. newlyweds used to regularly visit the parents of the bride or groom during their first year of marriage).

However,

> the Lacandones often make long journeys on foot to the shrines of their various gods. The women accompany the men on these trips, often carrying a child astride the hips, together with another swung on a net on the back. (Tozzer, 1907, p. 24)

In the past, they used to change settlements to avoid epidemics or in search of better lands for agriculture or hunting. Following the nomadic cycle of their crops, every four or five years they moved on to a different place, but always in the same environment, thus enabling an optimal understanding of the rainforest and its resources. They also went on expeditions to other parts of the forest to hunt, fish or visit family and friends.

As regards other ancestral costumes, *k'uch*, cords made of cotton thread for the sick or parturient were a fundamental aspect of life. Married women prepared healing cords for family, friends and even a favourite dog by rolling the thread with their hands to create a cord that was then placed around the neck of the sick person or on an injured limb and left there until it fell off on its own accord. Only married women could do this because the rite was related to the power of the umbilical cord. The cords were not really for healing, but a symbol of solidarity (Boremanse, 1998). They involved a kind of spell that women cast to prevent evil from returning to them. If a woman knew that someone was sick and did not give him or her a cord, and her negligence contributed to the death of the patient, the soul of the deceased could return to seek revenge on her (McGee, 2002). Since women did not place these cords on members of their own household, if a Lacandon woman knew of another person's illness or pregnancy, she or her husband had to visit the patient or pregnant women to offer them a cord. Paying visits formed part of a more complete trip for hunting, searching for materials or other purposes.

Until the 1970s, a very strict ritual was followed during Lacandon visits, above all in the north. Boremanse (1998) offers us the following description:

> The visitors who have come from a different settlement stop about forty meters from the first household of the group they intend to visit. They hide their rifles (in the past their bows and arrows) and net bags in the bushes by the trail. (…) One of the travelers then hits a tree trunk with his machete (…) to call the attention of the people who are going to be visited. (…) A little while after hearing the signal, the eldest man of the cluster, or the head of the household that is situated nearest the trail (…) goes to see who has come, but without any hurry. (…) After they have announced themselves, (…) all the members of the party hide in the bushes. (pp. 18–19).

The head of the household then sat by the trail and struck up a conversation with the visitors' spokesperson, while the rest of the group remained hidden in the undergrowth. Finally, after a brief chat, the host formally invited the travellers to enter his house, taking his leave to warn the womenfolk of the visit and to make the necessary arrangements. Five minutes later, the guests followed him, leaving behind their belongings, especially their rifles or bows and arrows, hidden in the bushes on the roadside, and carrying only their leather bags with personal items. Once inside the house, the visitors settled near the door or in a separate cabin, from where they rarely moved (Boremanse, 1998, p. 21). They only spoke with the host and his wife, while always avoiding eye contact. At a certain moment, visiting women got up to tie a *k'uch* on the right wrist of a new-born baby or someone sick. Other members of the household or other households in the family cluster did not participate in the visit, even if they were close relatives or friends. And if the visitors were still there at nightfall, they went to fetch their belongings and upon returning handed their weapons over to their host.

As to sleeping arrangements, visitors remained in the same area, near the house's threshold. The men continued chatting for quite a time, while the women and children rested. The conversation continued to be very formal, until late into the night when it became more relaxed. During the visit, the host offered his visitors tobacco and tortillas for dinner. If they wanted to purchase products not grown in their own fields, the host refused payment or offered, in exchange for money, a larger amount of the product. 'This typical behavior denotes a practice of gifts and counter-gifts that does not occur outside the context of hospitality' (Boremanse, 1998, p. 21).

In contrast, the Southern Lacandon Indians did not need any pretext to visit neighbours or friends. Upon arrival, they whistled and their hosts received them without further ado, offering them something to drink and often a full dinner. These practices were reciprocal, guests inviting their hosts to return the visit as soon as it was a good time to hunt.

Stage two: in situ exposure

This stage would be represented by the first expeditions of explorers, anthropologists and scientists into unknown lands, motivated by curiosity or economic and military interests. They left long accounts of the culture of these groups for fear that they would disappear in the name of 'progress' (Weaver, 2010).

Until the mid-twentieth century, the Lacandon Jungle was a practically uninhabited and very isolated place, without paved roads or electricity. To reach the Lacandon communities it was necessary to cover long distances on horseback, by boat along the Usumacinta River or by plane followed by hard trekking through the jungle. Even so, loggers, rubber tappers and numerous researchers and adventurers journeyed into the rainforest in search of riches or knowledge.

Through these contacts, the Hach Winik soon began to understand the rules of hospitality and exchange with strangers. Since the first academic expeditions of Tozzer, Soustelle and Blom, the discovery of the Bonampak remains and the arrival of the first allocentric tourists, many indigenous families in all communities began to provide researchers and travellers with food and shelter. But these first initiatives, although confirming the 'warm and generous' nature of Lacandon hospitality (Boremanse, 1998, p. 24), cannot yet be considered tourist activities. Weaver (2010, p. 47) calls this stage 'in situ exposure' because in these first encounters indigenous people were the object of curious observation by Westerners. According to Hinch and Butler's matrix (2007), the theme is still clearly indigenous but control begins to be more transitional, inadvertently passing into non-indigenous hands.

Stage three: ex situ exhibitionism and exploitation

According to Weaver (2010), this stage is characterised by the exhibitionism and exploitation of the natives outside their territory, with a clear interest in highlighting their savage character in opposition to Western civilisation. The early accounts of researchers and explorers aroused the curiosity of their compatriots, thus fostering the development of indigenous exhibitions and museums displaying native artefacts and handicrafts in North America and Europe. Their interpretation reinforced the 'moral superiority' of Eurocentric views, portraying indigenous cultures as backward. The dehumanisation of indigenous peoples was even taken a step further with the exhibition of individuals at fairs all over the world.

For the Lacandon people, this is the stage when legend was forged. The stories of the first explorers, together with the ethnographic works of the first half of the twentieth century, breathed life into the myth of continuity between the ancient Mayan civilisations and the recently rediscovered Lacandon, turning them into 'the last lords of Palenque'. According to these early accounts, the Lacandon had managed to eschew the Spanish conquest thanks to the isolation provided by the rainforest. The essence of the noble savage myth and Eurocentric exoticism found an ideal breeding ground in this group. However, Trench (2005, p. 49) emphasises the contradictions of this period in which, for example, Franz and Trudy Blom described the Lacandon as 'lords of an immense kingdom', while treating them like 'savages'.

For his part, Perera recalls his first contact with the Lacandon in 1938 (Perera & Bruce, 1985), when his father took him to a fair organised by the President of Guatemala, General Jorge Ubico, who wanted representatives of all the Mayan tribes exhibited at it.

> In the middle of this enclosure, behind a barbed-wire fence guarded by soldiers, a group of five Lacandon Indians was housed in a small hut of bamboo and palm thatch intended to duplicate their jungle home. They wore long white tunics and had shoulder-length matted hair, so it was difficult to tell if they were men or women. (…) One of the men came out of the hut and walked right to the fence where we stood. He made a noise in his throat and stuck his hand through the fence. Before I could overcome my bewilderment and decided whether to offer him a coin or the piece of chocolate I had in my pocket, a soldier pulled him away. The look of sadness and confusion in his eyes will stay with me forever (Perera & Bruce, 1985, p. 36).

These Lacandon Indians had been deceived and illegally abducted to satisfy the curiosity of the Guatemalan elites.[2]

The representations of the Lacandon as 'savages' are as plentiful as the descriptions of these people as the principle heirs to the Mayan culture. Pereira remembers his upbringing in Guatemala, where the 'savagery' of the Lacandon was used to scare children: 'Eat all your beans or the Lacandon will get you.' Others mocked their primitive way of life by referring to them as 'one of those who live like a monkey, in the jungles of El Petén' (Perera & Bruce, 1985, p. 37).

Like elsewhere, the first images of the Lacandon served for the 'appropriation and exploitation of indigenous culture to market the broader tourism industry, as seen in the prominence of stereotyped postcard images' (Weaver, 2010, p. 49), thus contributing to turn the Lacandon into the iconic symbol of Chiapas indigenousness.

> Newly arrived tourists to the jungle have been exposed to enough images of the Lacandon on television or in the newspapers to easily recognize the 'trademark' of this small indigenous group: their long hair and white robes (Trench, 2005, p. 48).

Therefore this stage implies a clear loss of control, where the indigenous theme is still present but clearly manipulated in the interests of the colonising group (Weaver, 2010).

Stage four: in situ exhibitionism and exploitation

With the arrival of mass transportation, peripheral areas are no longer isolated and a more articulated system of tourism can be introduced into indigenous territories. The first organised visits consist of staged representations of the native culture, where the interaction between tourists and locals is generally limited to buying cheap handicrafts and taking photographs. These activities are always organised by intermediaries outside the community who establish the rules and reap the benefits. The loss of control is absolute in this stage and the exhibitionism and exploitation of indigenous peoples happens in situ. However, it is also in this context that the first actions of resistance begin to take shape. In situ contact allows indigenous people to get to know and understand tourists, while these are still ignorant of the local culture, thus giving them power over the relationship and generating opportunities for control and resistance (Weaver, 2010).

The evolution of Lacandon tourism differs slightly from that presented by Weaver insofar as outside interest in 'ethnic art' and 'pan-Indian art' had an impact on Lacandon lives before the tourists themselves. In the Lacandon rainforest, the economic relationship between the natives and mass tourism began while jungle trails were still impassable for mass transportation. The daily arrival of small planes at Bonampak in the 1970s (Boremanse, 1998), as well as an increase in cultural tourism in Palenque in the following decade (McGee, 2002), opened up a market for Lacandon handicrafts. Men began to make bows and arrows expressly for the tourist market, while women saw in the traditional manufacture of necklaces and seed bracelets a source of monetary income. For the Lacandon, this development coincided with a wave of immigration and greater contact with the outside world. Their subsistence and exchange economy became insufficient and the need for hard cash pressing. In 1991, Marion Singer wrote that 'the Lacandon of Nahá and Metzabok could hardly meet all their new needs without the help of tourism' (p. 232). In Lacanjá Chansayab, despite the diversification measures implemented by the Government, Boremanse (1998, p. 9) states that 'in 1990, many of them [Lacandon] still relied on the sale of bows and arrows to tourists as a source of cash.'

In the early 1990s, notwithstanding the growing dependence on the tourist industry, the limited road infrastructure continued to prevent tourists from reaching the heart of the jungle and, therefore, the relationship of the Lacandon with tourism remained fundamentally ex situ. During this period, Nahá and Metzabok still received the same type of visitors as in the previous stage, i.e. adventurers and academics. In Lacanjá Chansayab, the tourists arriving by plane to visit the Bonampak archaeological site only stayed a few hours, their contact with the natives being restricted to the transport service from the airstrip to the site and to some hastily taken snapshots.

Despite the daytrip character of these visitors, the contacts developed in Bonampak match the framework proposed by Weaver. Even though the commercial exchange was dominated by intermediaries, it led to the first attempts at control among the indigenous people.

> In the 1970s small planes transporting German, American, French, and Mexican tourist, returning from the Maya ruins of Bonampak or Yaxchilán, landed almost daily at Lacanha' Chan Sayab. On the airstrip they took pictures of Southern Lacandon men and women, who asked them for money for cigarettes and tried to sell them sets of bows and arrows as well as homemade necklaces. The Lacandon also poked fun at the tourists who were usually disappointed (Boremanse, 1998, p. 10).

The Lacandon mockery of tourists recalls the examples studied by Evans-Pritchard (1989) among the Pueblo and Navajo Indians of New Mexico, where 'indigenous people maintained their dignity by caricaturing visiting tourists, often with the latter being unaware of the parody' (Weaver, 2010, p. 54). The contacts in the jungle gave the Lacandon an opportunity to learn about international travellers, thus opening up opportunities for control and resistance.

Stage five: in situ quasi-empowerment

For Weaver, the beginning of the fifth stage is marked by the emergence of community-based tourism and ecotourism models, through which native groups begin to exercise 'formal control over tourism within indigenous areas' (2010, p. 49) by determining sacred spaces, restricting access to their lands and participating actively in the design of tourism development programmes and cultural products. These dynamics are supported by new policies and funding at a national/regional and international level, as well as the advent of indigenous tourism associations.

This is the case with the Lacandon people. Funded by different institutions at all government levels, community-based tourism and ecotourism have been steadily promoted in the area over the past 20 years (López & Palomino, 2012; Palomino & López, 2007; Pastor-Alfonso et al., 2012). Through these public programmes run in conjunction with local initiatives and supported by NGOs and universities, the Lacandon have managed to consolidate solid tourism offerings that have replaced traditional economic activities for many families (Pastor-Alfonso, 2012).

As mentioned previously, the fourth stage of Weaver's model is characterised by the 'diffusion of mass transportation networks into peripheral areas' (Weaver, 2010, p. 49) and the consequent arrival of mass tourism to indigenous lands, while the fifth is identified by a greater degree of empowerment, the recovery of indigenous rights and the subsequent control of tourism activities.

In the Lacandon Jungle, both phenomena occurred simultaneously, uniting the fourth and fifth stages of Weaver's model in one. Mass transportation and, therefore, the Lacandon's real participation in the tourism industry started in 1997 when President Ernesto Zedillo Ponce de León officially opened the southern border highway, between Palenque and Montebello Lakes National Park, connecting Lacanjá Chansayab with the main tourist hubs. By facilitating access to the rainforest, the destination was suddenly opened up to the national and international tourism markets. However, the decision to open the highway was really a counter-insurgency measure to control the Zapatista indigenous uprising. Therefore, the real development of tourism in Lacandon territory took place against a backdrop of struggle for indigenous peoples' human rights. Moreover, the indirect publicity generated by the Zapatista movement, especially among Europeans, became a driving force behind tourism development in the area (Reygadas et al., 2006, p. 83).

Between 1994 and 2003, new roads between different jungle communities were built; however, Nahá and Metzabok continued to be connected by dirt roads, particularly unpredictable in the rainy season,[3] thus safeguarding the Northern Lacandon destinations from a tourist invasion.

Despite the control over tourism exercised by local groups, together with the land rights granted to the Lacandon in 1972, we cannot yet speak of real empowerment. Weaver calls it 'quasi-empowerment' to reflect the existence of other problems that, as Johnston (2006) rightly points out, still exist: corruption, paternalism, dependency, mismanagement, exploitation, etc. 'A long history of alliances with and subordination to state agencies has made the Lacandon a very dependent group habituated to a patronage relationship with external institutions' (Reygadas et al., 2006, p. 84).

Stage six: ex situ quasi-empowerment

According to Weaver's model, the sixth and current evolutionary stage would be characterised by the expansion of the geopolitical, economic and cultural influence of indigenous people outside their own sphere, i.e. ex-situ control. This stage involves the restitution of land rights, the legal and academic recognition of traditional knowledge, a growing respect for indigenous culture and values, equitable benefit sharing, civil rights and a political voice at local and national levels.

These rights, although ratified by the United Nations, are constantly in conflict with the interests of the dominant culture, above all as regards natural parks and other protected areas in which it has a limited stake and the presence of indigenous groups is high. Solutions to conflicting values and objectives usually take the shape of 'joint management' and 'co-management' schemes.

The political situation of the Lacandon is exceptional when compared to other indigenous groups. The restitution of communal lands, among other privileges, was granted to them in 1971 (de Vos, 2002). This political move was premeditated and the underlying interests had very little to do with indigenous well-being (Nations, 1979; Trench, 2008). However, it coincided with a growing international awareness of the marginalised situation of indigenous peoples around the globe (Martínez Cobo, 1986), 36 years before the UN Declaration on the Rights of Indigenous Peoples.

In light of such an exceptional situation, some of the characteristics of the sixth stage are far from being applicable to the Lacandon, accustomed to exercising their rights (and even their privileges) for more than 40 years. Nevertheless, as already noted, this close relationship with the Government has led to a serious patronage problem.

In addition, 'advances in terms of environmental governance have also been contradictory' (Reygadas et al., 2006, p. 83). Despite the use of participatory methods in the design of the Montes Azules Biosphere Reserve (REBIMA) management plan, Reyagadas et al. (2006, p. 83) note that, in practice, local participation is merely formal. In fact, the reserve has a highly bureaucratic vertical management structure. In recent years, significant improvements have been observed, 'participation in REBIMA has been strengthened and there is undoubtedly a genuine interest of CONANP [national parks authority] and its staff to integrate and consider the views of communities in the management of the reserve' (Durand et al., 2012, p. 264). However, the endless red tape, the inequalities in each indigenous group and between groups, and the lack of political voice all mean that 'participation in REBIMA and other nearby protected areas such as MN Bonampak and Yaxchilán (…) has not yet managed to ensure the inclusion of all the affected or interested parties' (p. 265).

On the other hand, Weaver observes that in this sixth stage there is a greater indigenous presence in typical Western market economy activities, what Hinch and Butler (2007) call 'diversified indigenous' activities where there is control, but the indigenous theme disappears. Although it is possible to observe a certain disconnection between indigenous cultural themes and the tourism offering in the Lacandon Jungle, from the authors' point of view, this is not an empowerment outcome, but rather the effect of outside interests.

Product 'diversification' is mainly a result of joint-ventures with non-indigenous capital and entrepreneurship, in step with national tourism strategies that are encouraging adventure tourism over the cultural or 'dual track' kind. Some examples are the kayak activities offered by the *Rio Lacanjá Camp* in association with the company *Explora, ecotourism and adventure* in San Cristóbal de las Casas, and the buggies, kayaks and rafting activities recently promoted by *Indigenous Treasures* and managed by the company *VDP Consultores,* both non-indigenous organisations based in Mexico City.

So, instead of cases of 'shadow indigenous tourism,' there are situations of 'quasi-empowerment' more characteristic of the previous stage. However, through collaborative research strategies with local entrepreneurs, a strong interest in cultural revival has been identified (Espeso-Molinero et al., 2016). As has happened elsewhere (Esman, 1984; Henderson, 2003; Smith, 1989), tourism has increased the self-esteem of the Lacandon and their appreciation of the traditional values that make them unique (Pastor-Alfonso & Gómez López, 2010; Pastor-Alfonso, 2012), as well as leading to a revival of local crafts (McGee, 2002). Nonetheless, contrary to the examples presented by McAvoy (2002), Pretes (2002) and Pitchford (2006), the scant presence of cultural products and activities are limiting the possibilities of the Lacandon to offer a less stereotyped and essentialist image of their culture.

In terms of benefit sharing and equity of treatment, Reygadas et al. (2006) observes that in Lacanjá Chansayab the different government programmes have only favoured a very limited group of families, and within these there are important imbalances between women and men. Most of the projects are controlled by men aged between 25 and 45 years old, who also monopolise the main public jobs. 'Far from a truly communal appropriation of tourism and archaeological income, a pattern of family or private monopolisation of natural and tourist resources has prevailed' (p. 84). Pastor-Alfonso et al. (2012) point out that the participation of women in community groups is often merely a strategy to attract funding. For Valle-García (2014), alternative tourism has not achieved sustainable objectives in Lacanjá Chansayab. In fact, top-down strategies and a lack of community participation in tourism development has led to inter- and intra-ethnic conflicts, as well as marginalisation.

Even though recent collaborative studies with Lacandon entrepreneurs show an increasingly greater presence of women in traditional male roles (guides, managers, etc.), a growing interest in cultural revival and a more balanced distribution of power among neighbouring groups, tourism in the Lacandon Jungle is still far from a situation of 'ex situ quasi-empowerment'.

Conclusions

Notwithstanding the existence of several chronological singularities and variations, it can be said that the evolution of the relationship between tourism and the Lacandon mostly follows the stages described in Weaver's model (2010). However, in order to apply this model to indigenous peoples in Mexico or other Latin American countries it is essential to consider certain crucial differences.

Pre-colonial history has left a strong imprint on the majority of Latin American countries where many indigenous communities have gone through 'colonial' processes under the rule of pre-Hispanic empires. Besides archaeological sites and artistic remains, those processes have left behind an important cultural and social legacy. Therefore, any analysis of the evolution of tourism in Latin America should bear in mind pre-Hispanic history.

The four countries analysed by Weaver have attained 'economically developed status,' indigenous tourism has evolved 'within well-articulated and sophisticated domestic and international tourism systems' and there is a 'relatively long participation and the increasing visibility of indigenous culture and peoples within these tourism systems' (Weaver, 2010, p. 46). Conversely, the UN regards most Latin America countries as developing economies; although their tourism systems are improving, they are not yet fully articulated and indigenous tourism initiatives are comparatively new.

As in Weaver' cases, the indigenous peoples of Latin America are suffering marginalisation, discrimination and hardship. However, in Guatemala, Bolivia and Ecuador, for instance, they are not 'ethnic minorities,' but quite the opposite. Furthermore, the presence of a large mestizo society prevents a 'clear formal distinction between "indigenous" and "nonindigenous" residents' (Weaver, 2010, p. 49), the collection of statistical data often being a difficult task (UNPII, 2004).

An analysis of the indigenous peoples of Latin America must also take into consideration the role of post-colonial governments, since in many cases, such as Mexico, comprehensive assimilation policies have been implemented in order to reinforce the national identity (Bonfil Batalla, 1989).

In the Lacandon case, the lack of mass transportation networks and the recent development of indigenous tourism has led to a situation in which the fourth and fifth stages have run concurrently. And, as in many Latin American countries, even without the in situ presence of tourists due to transportation problems, the Lacandon have benefitted from trading opportunities in handicrafts providing with much needed hard cash. Even though the characteristics of the sixth stage have yet to materialise, the outlook for ex situ quasi-empowerment is promising.

Despite the fact that Weaver doubts the convenience of including the first stage inasmuch as pre-European travel cannot be conventionally thought of as a form of indigenous tourism, in this case study it was characterised by the Lacandon's complex visiting rules and rituals. A better understanding of indigenous peoples' traditional hospitality practices could give us important clues as to the development of a less Eurocentric indigenous tourism industry focusing more on local values.

In light of the above, Weaver's model opens up interesting opportunities for further analysis in the field of indigenous tourism in Latin America and other developing regions of the world.

Notes

1. Marie-Odile Marion (1999, p. 10) recovers the term 'usurpers', used by the Maya in the *Chilam Balam* books, to designate the Spanish conquerors.
2. Against the wishes of Major Bickford, who intended to establish a permanent exhibition featuring the Lacandon, Mario Monteforte Toledo, leader of the adventurers who had taken them to Guatemala, kept his promise to return them to the jungle after the fair. During their stay in Guatemala, they all suffered from lung problems and gastroenteritis. One of the five never returned.
3. The soil of the Lacandon rainforest is composed of sedimentary rocks from the Cretaceous period, giving rise to relatively impermeable subsoil which favours runoff and soil erosion (Eroza Solana, 2006). In the rainy season, the rainforest's dirt roads are often blocked by landslides.

Disclosure statement

No potential conflict of interest was reported by the authors.

ORCID

Pilar Espeso-Molinero http://orcid.org/0000-0001-5876-3906
María José Pastor-Alfonso http://orcid.org/0000-0001-8820-8073

References

Baer, P., & Merrifield, W. R. (1972). *Los lacandones de México: Dos estudios* [Two studies of the Lacandones of Mexico] (C. Viqueira Trans.). Instituto Nacional Indigenista.

Blom, F. (1954). Ossuaries, cremation and secondary burials among the Maya of Chiapas, Mexico. *Journal De La Société Des Américanistes*, *43*(1), 123–136.

Bonfil Batalla, G. (1989). *México profundo: Una civilización negada* (2nd ed.). México, D.F.: Grijalbo.

Boremanse, D. (1989). Ortogénesis en la literatura maya lacandona. *Mesoamérica*, *10*(17), 61–104.

Boremanse, D. (1998). *Hach winik: The Lacandon Maya of Chiapas, Southern Mexico*. New York, NY: University Press of Colorado.

Boremanse, D. (2006). *Cuentos y mitología de los lacandones: Contribución al estudio de la tradición oral maya*. Guatemala: Academia de Geografía e Historia de Guatemala.

Boremanse, D. (2007). K'in yah: Divination ritual in lacandon maya religion. *Mesoamérica*, *28*(49), 114–135.

Bruce, R. (1968). *Gramática del lacandón*. México: Instituto Nacional de Antropología e Historia.

Bruce, R. (1976). The popol vuh and the book of chan k'in. *Estudios de Cultura Maya*, *10*, 173–209.

Bruce, R. (1979). *Lacandon dream symbolism: Dream symbolism and interpretation among the Lacandon Mayas of Chiapas, Mexico*. Mexico, D.F.: Editiones Euroamericanas.

Butler, R., & Hinch, T. (1996). *Tourism and indigenous peoples* (1st ed). London: International Thompson Business Press.

Buultjens, J., & Fuller, D. (eds.). (2007). *Striving for sustainability: Case studies in indigenous tourism*. Lismore, N.S.W.: Southern Cross University Press.

Calleros-Rodríguez, H. (2014). Land, conflict, and political process: The case of the Lacandon community, Chiapas, Mexico (1972-2012). *Journal of Peasant Studies*, *41*(1), 127–155.

Carr, A., Ruhanen, L., & Whitford, M. M. (2016). Indigenous peoples and tourism: The challenges and opportunities for sustainable tourism. *Journal of Sustainable Tourism*, *24*(8–9), 1067–1079.

de Vos, J. (1980). *La paz de dios y del rey: La conquista de la selva lacandona 1525-1821*. México, DF: Fonapas/Chiapas.

de Vos, J. (2002). *Una tierra para sembrar sueños: Historia reciente de la Selva Lacandona*. México: Centro de Investigaciones y Estudios Superiores en Antropología Social.

Diemont, S., & Martin, J. (2009). Lacandon Maya ecosystem management: Sustainable design for subsistence and environmental restoration. *Ecological Applications*, *19*(1), 254–266.

Durand, L., Figueroa, F., & Trench, T. (2012). Inclusión, exclusión y estrategias de participación en la reserva de la biosfera montes azules. In L. Durand, F. Figueroa, & M. Guzmán (Eds.), *La naturaleza en contexto: Hacia una ecología política mexicana* (pp. 237–267). México, D.F.: Universidad Nacional Autónoma de México.

Eroza Solana, E. (2006). *Lacandones (Pueblos indígenas del México contemporáneo)*. México, D.F.: Comisión Nacional para el Desarrollo de los Pueblos Indígenas.

Esman, M. R. (1984). Tourism as ethnic preservation: The Cajuns of Louisiana. *Annals of Tourism Research*, *11*(3), 451–467.

Espeso-Molinero, P. (2018). Collaborative capacity building as a resilience strategy for tourism development in indigenous Mexico. In J. M. Cheer, & A. A. Lew (Eds.), *Tourism, resilience, and sustainability: Adapting to social, political and economic change* (pp. 184–201). London: Routledge.

Espeso-Molinero, P., Carlisle, S., & Pastor-Alfonso, M. J. (2016). Knowledge dialogue through indigenous tourism product design: A collaborative research process with the Lacandon of Chiapas, Mexico. *Journal of Sustainable Tourism*, *24*(8–9), 1331–1349.

Evans-Pritchard, D. (1989). How "they" see "us": Native American images of tourists. *Annals of Tourism Research*, *16*, 89–105.

Henderson, J. (2003). Ethnic heritage as a tourist attraction: The Peranakans of Singapore. *International Journal of Heritage Studies*, *9*(1), 27–44.

Hernandez Cruz, R. E., Bello-Baltazar, E., Montoya Gómez, G., & Estrada Lugo, E. I. J. (2005). Social adaptation: Ecotourism in the Lacandon forest. *Annals of Tourism Research*, *32*(3), 610–627.

Hinch, T., & Butler, R. (2007). Introduction: Revisiting common ground. In R. Butler, & T. Hinch (Eds.), *Tourism and indigenous peoples: Issues and implications* (pp. 1–12). Oxford: Butterworth-Heinemann.

Johnston, A. M. (2006). *Is the sacred for sale? Tourism and indigenous peoples*. London: Earthscan.

Kashanipour, R. A., & McGee, R. J. (2004). Northern Lacandon Maya medicinal plant use in the communities of Lacanjá Chan Sayab and Naha', Chiapas, Mexico. *Journal of Ecological Anthropology*, *8*, 47–66.

Lemelin, R. H., & Blangy, S. (2009). Introduction to the special issue on aboriginal ecotourism. *Journal of Ecotourism*, *8* (2), 77–81.

López, G., & Palomino, B. (2008). Políticas públicas y ecoturismo en comunidades indígenas de México. *Teoría Y Praxis*, (5), 33–50.

López, G., & Palomino, B. (2012). *Guía de apoyos federales para el desarrollo del turismo de naturaleza en México. (Informe de resultados)*. México, D.F: SECTUR-CONACYT.

Marion Singer, M. (1991). *Los hombres de la selva: Un estudio de tecnología cultural en medio selvático*. México, D.F.: Instituto Nacional de Antropología e Historia.

Marion Singer, M. (1999). *El poder de las hijas de luna: Sistema simbólico y organización social de los lacandones*. México, DF: Plaza y Valdés.

Marion Singer, M. (2000). Bajo la sombra de la gran ceiba: La cosmovisión de los lacandones. *Desacatos. Revista De Antropología Social*, (5), 45–56.

Martínez Cobo, J. R. (1986). *Study of the problem of discrimination against indigenous populations. (No. UN-Document E/CN. 4 Sub.)*. New York, NY: United Nations.

McAvoy, L. (2002). American Indians, place meanings and the old/new west. *Journal of Leisure Research*, *34*(4), 383–396.

McGee, R. J. (1990). *Life, ritual, and religion among the Lacandon Maya. Belmont*. California: Wadsworth Publishing Company

McGee, R. J. (2002). *Watching Lacandon Maya lives*. Boston: Allyn and Bacon.

McGee, R. J., & González, B. (1999). Economics, women, and work in the Lacandon Jungle. *Frontiers*, *20*(2), 175–189.

McIntosh, A. J., & Ryan, C. (2007). The market perspective of indigenous tourism: Opportunities for business development. In R. Butler, & T. Hinch (Eds.), *Tourism and indigenous peoples* (pp. 73–83). Oxford: Butterworth-Heinemann.

McIntosh, A. J., Zygadlo, F. K., & Matunga, H. (2004). Rethinking Maori tourism. *Asia Pacific Journal of Tourism Research*, *9*(4), 331–352.

Nations, J. (1979). *Population ecology on the Lacandon Maya*. Southern Methodist University.

Nations, J., & Clark, J. (1983). The bows and arrows of the Lacandon Maya. *Archaeology*, *36*(1), 36–43.

Palka, J. W. (2005). *Unconquered Lacandon Maya: Ethnohistory and archaeology of indigenous culture change.* Gainesville: University Press of Florida.

Palomino, B., & López, G. (2007). *Evaluación 2006 del programa ecoturismo en zonas indígenas.* (Informe Técnico).Comisión Nacional para el Desarrollo de los Pueblos Indígenas. Retrieved from http://ru.iiec.unam.mx/2370/

Pastor-Alfonso, M. J. (2012). Turismo y cambio en el entorno de los lacandones. Chiapas, México. *Pasos. Revista De Turismo Y Patrimonio Cultural, 10*(1), 99–107.

Pastor-Alfonso, M. J., & Gómez López, D. (2010). *Impactos socioculturales en el turismo comunitario. Una visión desde los pueblos implicados (Selva Lacandona, Chiapas, México).* Editorial Agua Clara: Salamanca.

Pastor-Alfonso, M. J., Gómez López, D., & Espeso-Molinero, P. (2012). Turismo comunitario y sus consecuencias entre los lacandones de Chiapas: Organismos y sistemas de apoyo. In A. Santana, A. J. Rodríguez Darias, & P. Díaz Rodríguez (Eds.), *Responsabilidad y turismo* (pp. 23–43). Tenerife: PASOS, Revista de Turismo y Patrimonio Cultural. Retrieved from http://www.pasosonline.org/Paginas/e-book_10.htm

Perera, V., & Bruce, R. (1985). *The last lords of Palenque: The Lacandon Mayas of the Mexican rain forest.* Berkeley, CA: University of California Press.

Pitchford, S. R. (2006). Identity tourism: A medium for Native American stories. *Tourism, Culture and Communication, 6*(2), 85–105.

Pretes, M. (2002). Touring mines and mining tourists. *Annals of Tourism Research, 29*(2), 439–456.

Reygadas, L., Ramos, T., Montoya Gómez, G., Hernández, F., & Velasco, A. (2006). Estilos de manejo y gestión de proyectos ecoturísticos en la selva lacandona de Chiapas, México. In R. Guevara Ramos (Ed.), *Estudios multidisciplinarios en turismo.* volumen 1 (pp. 71–102). México: Secretaria de Turismo, Red de Investigadores y Centros de Investigación en Turismo. Retrieved from http://bibliotecas.umar.mx/publicaciones/Estudios_versi%C3%B3n%20electr%C3%B3nica.pdf#page = 72

Ryan, C., & Aicken, M. (2005). *Indigenous tourism: The commodification and management of culture.* Amsterdam: Elsevier.

Smith, V. L. (1977). *Hosts and guests: The anthropology of tourism.* Philadelphia, PA: University of Pennsylvania Press.

Smith, V. L. (1989). Eskimo tourism: Micro-models and marginal men. In V. L. Smith (Ed.), (2nd ed., pp. 55–82). Philadelphia, PA: University of Pennsylvania Press.

Soustelle, J. (1935/2008). En la selva de Chiapas con los lacandones. *Históricas: Boletín Del Instituto De Investigaciones Históricas, UNAM,* (83), 14–19.

Tozzer, A. M. (1907). *A comparative study of the Mayas and the Lacandones.* New York, NY: The Archaeological Institute of America. The MacMillan Company.

Trench, T. (2005). Representaciones y sus impactos: El caso de los lacandones en la Selva Lacandona. *Liminar. Estudios Sociales y Humanísticos, 3*(2), 48–69.

Trench, T. (2008). From "orphans of the state" to the comunidad conservacionista institucional: The case of the Lacandon community, Chiapas. *Identities, 15*(5), 607–634.

United Nations Secretariat, Permanent Forum on Indigenous Issues (UNPFII). (2004). 'World Bank Contribution to the Expert Workshop on data collection and disaggregation for Indigenous Peoples'. PFII/2004/WS.1/11.

Valle-García, S. (2014). Ecotourism: Sustainable indigenous policies and its effects in Mayan communities, southern Mexico. In C. A. Brebbia, S. Favro, & F. D. Pineda (Eds.), *Sustainable tourism VI* (pp. 239–250). Southampton: WIT Press.

Weaver, D. B. (2010). Indigenous tourism stages and their implications for sustainability. *Journal of Sustainable Tourism, 18*(1), 43–60.

Cultural sustainability – a framework for Aboriginal tourism in British Columbia

Tatjana Thimm

ABSTRACT

This paper presents a framework to assess the cultural sustainability of Aboriginal tourism in British Columbia, which meets must take into account the protection of human rights, good self-governance, identity, control of land, the tourism product's authenticity, and a market-ready tourism product. These criteria are specified by two indicators each. The cultural sustainability framework was generated by triangulating qualitative research methods like experts' interviews, secondary research, and participant and non-participant observations. This paper is thus conceptual in nature and inductive in its approach. It partly leverages a collaborative approach, as it includes interviewees in an iterative research loop. Furthermore, the paper shows why cultural sustainability is a determinant of the success of Aboriginal tourism.

Introduction

Cultural sustainability is not easy to define or operationalize, and concepts vary from author to author (Empacher & Wehlung, 2002). Aboriginal tourism is linked to this concept, as this form of tourism is largely cultural in nature. Tourists choose Aboriginal tourism packages because they are curious about the ancient or contemporary ways of life of the Aboriginal Peoples (Williams & O'Neil, 2007). In order for Aboriginal tourism to be marketed as a product, it has to be both market-ready and authentic, meaning that Aboriginal tourism faces particular challenges as it tries to manage the tensions between commodification and cultural preservation (Dorsey, Steeves, & Porras, 2004; Jamal & Stronza, 2008), staged authenticity and authenticity, and commercial success versus sustainable development (Axelsson et al., 2013; Schmiechen, 2006). There may also be negative effects (e.g. acculturation, the abandonment of traditional activities) and negative demonstrations that result from tourists' behaviours and consumption patterns. The cultural and economic permanence of Aboriginal tourism is often vulnerable due to a lack of education, training, business and management techniques, difficulties obtaining control over image, and land-related issues or its poor accessibility (Butler & Hinch, 2007a). Cultural sustainability, which was developed as a framework, helps to handle these challenges, and it can be tailored to the specific circumstances and requirements of the Aboriginal Peoples, such as those in British Columbia.

Background

The Aboriginal Peoples of Canada comprise the First Nations, Inuit, and Métis (Indigenous and Northern Affairs (INAC), Government of Canada, 2017). There are several terms like Indian,

Aboriginal, Native, First Nation, or Tribal tourism that tend to be used interchangeably, often depending on geographical context (Hinch & Butler, 2007; Lemelin & Blangy, 2009; Weaver, 2016; Zeppel, 2006). In Canada, 'Aboriginal Tourism' is the term that is used by the Aboriginal Peoples of Canada themselves and will thus be used in this paper as well, although Weaver (2016) fundamentally questions the term due to the amalgamation and contemporary expressions of Aboriginal cultures. Sticking strictly to the United Nations (UN) definition, Aboriginal or Indigenous comprises the following aspects: self-identification; historical continuity; link to territories; a distinct social, economic, or political system; a distinct language, culture, and beliefs; and minority groups that are often marginalized and reproduce ancestral environments (United Nations (UN), 2016). This definition of Aboriginal is in line with the Aboriginal aspect of Aboriginal tourism used in this paper.

The theme of 'sustainability' in relation to Aboriginal People is mostly discussed in terms of Indigenous people and ecotourism (Fennell, 2008; Johnston, 2000; Weaver, 2010; Zeppel, 2006). Whitford and Ruhanen (2016) provide an overview of the history of Indigenous tourism from 1980 to 2014 – a framework for culturally sustainable tourism has not yet been developed.

Aboriginal tourism in Canada

The literature on Aboriginal tourism in Canada is not abundant and is sometimes outdated. There are papers that focus on the demand (Williams & Stewart, 1997) or marketing (Williams & Richter, 2002) of Aboriginal tourism. Additional works have focused on various regional approaches apart from those in British Columbia (see below), such as those in Alberta (Notzke, 2004), the western Arctic (Notzke, 1999), the eastern Arctic (Anderson, 1991), Nova Scotia (Lynch, Duinker, Sheeehan, & Chute, 2010), and Saskatchewan (Li, 2000), while other authors have looked specifically at special aspects like souvenir trade (Blundell, 1993) or research approach on Aboriginal Canadians (Stewart & Draper, 2009). Although cultural topics are covered as well (Boyd, 2001; Mason, 2004; Seale, 1996; Wall, 2008), they are not linked with the concept of cultural sustainability. The latest market report on Aboriginal tourism in Canada was provided by O'Neil et al. (2015). In fact, Aboriginal tourism is a growing sector in Canada's tourism economy. Stakeholders in this sector are confident that this will be the case in the future as well (O'Neil et al., 2015). O'Neil et al. (2015, p. 4) define Aboriginal tourism as:

> All tourism businesses majority owned, operated and/or controlled by First Nations, Métis or Inuit peoples that can demonstrate a connection and responsibility to the local Aboriginal community and traditional territory where operation resides.

Most Aboriginal tourism products in Canada include Aboriginal cultural experiences; therefore, O'Neil et al. (2015) render Aboriginal cultural tourism more precisely as an industry that:

> Meets the Aboriginal tourism criteria and in addition, a significant portion of the experience incorporates Aboriginal culture in a manner that is appropriate, respectful and true to the Aboriginal culture being portrayed. The authenticity is ensured through the active involvement of Aboriginal people in the development and delivery of the experience.

Businesses in the Aboriginal tourism industry in Canada are predominantly owned by the Aboriginal Peoples and include offers that fall along the tourism value chain. The domestic market provides a lion's share of demand, and there are also markets in the United States and Europe. The most important Aboriginal tourism source market in Canada has always been that of the respective province itself. Similarly, the strongest European source market for Aboriginal tourism is Germany. On a global scale, Aboriginal tourism is a niche market, and Canada has competitors in this realm, such as Aboriginal tourism industries in Australia, New Zealand, Mexico, and the United States (Butler & Hinch, 2007b; O'Neil et al., 2015; Ryan & Aicken, 2005).

Aboriginal tourism in British Columbia

In British Columbia, Aboriginal tourism plays an increasingly important role in the tourism sector. According to the Aboriginal Tourism Association of British Columbia (AtBC) (Williams, 2013), the number of visitors who wish to seek Aboriginal tourism experiences doubled from 2006 to 2010. Increased marketing by the AtBC and the fact that the 2010 Olympics were held in British Columbia are both factors that have significantly contributed to this trend (interview with Informants A, F). The AtBC began its work in 1997 with the aim of enabling the Aboriginal Peoples to participate in the economic benefits of tourism, while at the same time promoting and preserving Aboriginal culture (Williams, 2013). Aboriginal tourism activities that are supported by the AtBC focus on Vancouver and its surrounding areas, and the Thompson-Okanagan region. Since the regions further north in British Columbia (e.g. Northern British Columbia) or those further west (like Kootenay Rockies) did not develop the same amount of Aboriginal tourism activity, it was determined that accessibility is a decisive factor in Aboriginal tourism (Williams, 2013). The AtBC has also been working on embedding more Aboriginal tourism activities in British Columbia's and Canada's tourism strategies, and this is reflected in their respective marketing channels (AtBC, 2017; Williams, 2013). To realize its operational business, AtBC launched Aboriginal Travel Services (ATS, 2018). Destination British Columbia is not normally chosen for trips given the fact that Aboriginal tourism is offered; Aboriginal tourism serves as an 'add on' for those visitors travelling to British Columbia who are looking for a city or nature trip (Butler & Hinch 2007a). Therefore, converting as many domestic and international travellers as possible into actual Aboriginal tourism visitors is an important strategy (Williams, 2013). The Aboriginal tourism experiences that are especially of interest are related to 'heritage, outdoor adventures, festivals, handicraft and art, cuisine and sport' (Williams & O'Neil, 2007).

Nepal (2005) stresses the importance of ensuring that the Aboriginal Peoples oversee and implement tourism projects; one such example is the case of the Tl'azt'en in Northern British Columbia. The author elaborates that the quality of the tourism product should not jeopardize the cultural integrity of the Aboriginal Peoples, and that a lack of education, training, funding, and participation are common barriers to implementing tourism projects. Kutzner, Wright, and Stark (2009) examined the Tl'azt'en Nation as well and emphasized the necessity of understanding Aboriginal tourists' interests and to base the tourism experience on these interests, while aligning the activities with community needs.

In Northern British Columbia, tourism is one of the few options left for the Aboriginal Peoples in areas with high unemployment rates. Nicholson (1997) stresses the types of opportunities available for Aboriginal women in terms of tourism education and other employment offers. Business alternatives in remote areas – apart from tourism – can only be found in the logging, mining, and park-management industries (or, at least, park co-management), or in services that do not require a specific location, such as Web design (interview with Informants D, E, I, J; Notzke, 2006; Notzke, 1994).

Literature review and research questions

Cultural sustainability

Cultural sustainability is a rather new term in the sustainability discourse. It was introduced in 1995 by the World Commission on Culture and Development (WCCD) and is defined as 'inter- and intra-generational access to cultural resources' (WCCD, 1995 quoted in Axelsson et al., 2013, p. 217). Thus, this concept is younger than the concept of sustainability in general, but it is no less complex and overlaps with social sustainability (Murphy, 2012). Furthermore, an Agenda 21 for culture was coined by United Cities and Local Governments (UCLG) in 2009 (Soini & Birkeland, 2014). The importance of establishing a fourth pillar of sustainability – cultural sustainability – was stressed (Culture 21, 2011; Hawkes, 2001). Throsby (2003) is broadening the concept of cultural sustainability by incorporating six principles: material and non-material well-being, intergenerational equity, intra-generational equity, maintenance of diversity, precautionary principle, maintenance of cultural systems, and recognition of independence. Moreover, Soini and Birkeland (2014) analysed the scientific discourse on cultural sustainability and found that it is organized around seven storylines: heritage (e.g. Farsani, Coelho, & Costa, 2011; Vileniske, 2008), vitality (e.g. Kong, 2009; Trimarchi, 2004), economic viability (e.g. Farsani et al., 2011; Mccartney & Osti, 2007), diversity (e.g. Beckerman & Kopelowitz, 2008; Oh, Kim, & Jeong, 2005), locality (e.g. Choy, 2004; Elbakidze, Angelstam, Sandström, & Axelsson, 2010), eco-cultural resilience (e.g. Van Mansvelt, 1997; Von der Pahlen & Grinspoon, 2002), and eco-cultural civilization (e.g. Beddoe et al., 2009; Burton & Paragahawewa, 2011). Nassauer (2004) offers a more general approach towards cultural sustainability; for him, it consists of cultural values and behaviours that support ecology. In summary, cultural sustainability is still a recent and emerging concept and it is very broad in its interpretation; it has manifold interdependencies to social, economic, and ecological sustainability (Soini & Birkeland, 2014).

Cultural sustainability in tourism

The Global Sustainable Tourism Criteria (GSTC), developed as the minimum sustainability standards for tourism worldwide by the Global Sustainable Tourism Council (2016), comprise a chapter on cultural sustainability, which is by far the shortest one when compared to those written on social, economic, and ecological sustainability. It established different criteria sets for hotels, tour operators, and destinations. These criteria focus on heritage protection, codes of conduct for visitors, accurate site interpretation, and intellectual property rights.

The Global Sustainable Tourism Council (2016) also offers indicators for these criteria. Since neither the criteria nor the indicators of GSTC fit the specific situations faced by the Aboriginal Peoples, they can only be taken into account at a general level.

More Than Green (2016) stresses the fact that cultural sustainability is based on 'identity,' which includes both values and resources; identity is a very important aspect for Aboriginal tourism (see below). Cultural sustainability as a framework for tourism has not received much attention in the scientific discourse on Aboriginal Peoples. A first attempt at initiating scientific inquiry into this matter was made by Johnston (2006), who created a list of culturally sustainable tourism aspects that apply to the Aboriginal Peoples, which features guidance by elders, ancestral title and rights, and protocols approved by the Aboriginal Peoples on cultural diversity, collectivity, community, spirituality, and human rights, which are not necessarily in line with the typical management approaches of sustainable tourism adopted in the West. Dyer, Aberdeen, and Schuler (2003) stress that the control of cultural activities and intellectual property is decisive for the cultural sustainability of indigenous groups in tourism.

Another important contribution to this discussion is the study 'Indigenous Tourism and Human Rights' (Pacific Asia Travel Organisation [PATA], 2015); this investigation was based on several case studies and resulted in the development of a checklist for indigenous tourism. This checklist included aspects such as respect, protection, empowerment, consultation, business, and community, while accounting for the United Nations Declaration of the Rights of Indigenous Peoples (2007) and The Larrakia Declaration (2012). The PATA checklist contains several elements of cultural sustainability, which is directly related to human rights.

In British Columbia, the Aboriginal culture was threatened by colonization, assimilation, residential schooling, displacement to reserves, and an unsettled situation regarding land title. Furthermore, residential schooling – judged as cultural genocide – resulted in trauma and mistrust against the entire education system ('Canada's forced schooling,' 2015 quoted in World and Press 2 August 2015; PATA, 2015; interviews with Informants A, B, D, E, I), which is an obstacle that must be addressed when building capacity in tourism and when empowering the Aboriginal Peoples through education.

Thus, the following research questions were developed:

- Which criteria form a framework of cultural sustainability for Aboriginal tourism in British Columbia?
- How can this framework be used for Aboriginal tourism development in British Columbia?

Methodology

Culture-related tourism activities dominate the demands for Aboriginal tourism in Canada (O'Neil et al., 2015). These activities pose a competitive tourism advantage for Aboriginal Peoples (Hinch & Butler, 2009); therefore, an assessment of Aboriginal tourism on cultural sustainability should be based on this notion.

This project was funded by the author's university, and there were no ethical concerns expressed regarding the research proposal by the funding institution. However, the author is aware of the fact that 'researching the native' (Tuhiwai Smith, 2005) is a sensitive issue; therefore, the researcher involved the interviewees by delivering intermediate results and by providing them with ample room to ask all kinds of questions regarding the background and goal of the research. Taking into account that there are indigenous perspectives on research, e.g. expressed by Wilson (2001) as more fluid, depending on relationships and social context, the author considered this. These indigenous perspectives include storytelling and personal narratives (Besserab & Ng'andu, 2010) and are to a certain degree in line with constructivist theory and participatory action research (Kahakalau, 2004; Wilson, 2001). They follow more cyclic and circular idea (Louis, 2007). Furthermore, according to Wilson (2001) knowledge does not belong to an individual, it always must be shared. Taking the indigenous perspectives on research into account the author gave room for storytelling and personal narratives in the interviews, asked for feedback on intermediate results and shares the results of the research following the indigenous concept of reciprocity (Louis, 2007). Thus the author embedded an indigenous perspective as much as possible. In doing so the author always followed the concept of decentring (Godelier, 2017) in order to be open for the understanding of ideas and actions that are not familiar to the author's cultural background.

During a research stay in 2015 in British Columbia, the author conducted interviews with Aboriginal and non-Aboriginal experts on Aboriginal tourism from the tourism industry and academia, male and female alike (see a list of interviews below). Prior to the research stay, the author was supported by a female scholar from a Vancouver University and an Aboriginal Canadian working for the Aboriginal tourism industry, a key stakeholder with an ample network. Both assisted by providing contacts and information while preparing the research. Thus, a snowball sample of interviewees and locations was built and constantly expanded during the research stay. Data collection, which was achieved via semi-structured interviews, was complemented by participant and non-participant observation (Cole, 2005) at Aboriginal cultural tourism businesses like the Museum of Anthropology at the University of British Columbia, the Royal British Columbia Museum in Victoria, several Aboriginal Art Galleries in Gastown (Vancouver), the Cedar Root Gallery in East Hastings (Vancouver), the Musqueam Cultural Center Gallery (Vancouver), the Nk'Mip Cultural Desert Center in in Osoyoos, the Ktunaxa Interpretive Center in Cranbrook, the Skwachàys Lodge in Vancouver, the Bill Reid Gallery in Vancouver, the Museum of Vancouver, the Museum of Archeology and Ethnology at Simon Fraser University (Vancouver), and the Squamish Lil'wat Cultural Center in Whistler (see a list of participant observation protocols, below).

All interviews were anonymized and informed consent was obtained from the interviewees prior to recording and publishing the data. The interviews were analysed to identify patterns (i.e. answers that were repeated often). Furthermore, the interviewees were contacted a second and a third time, which served as an extra feedback loop to support the model of cultural sustainability (Figure 1) and

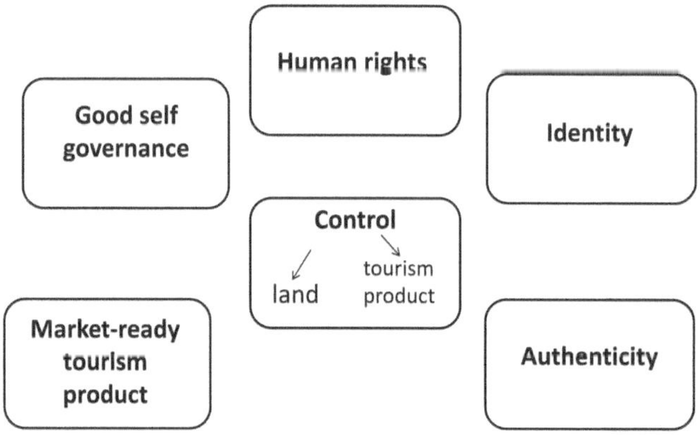

Figure 1. Model of cultural sustainability for Aboriginal tourism in British Columbia (author's illustration 2017).

the criteria and indicators (Table 1), while using a partly collaborative research approach (Lemelin & Blangy, 2009).

Finally, the criteria and indicators were formed using the information gathered through the interviews, as well as the participant and non-participant observations, and they were further complemented by secondary sources. Using this triangulation approach, the validity of the results is improved. Golafshani (2003) suggests replacing the term 'reliability' (which is more applicable to quantitative research) with 'dependability' and 'consistency.' With respect to dependability, the author holds available the raw data for possible inspection; concerning consistency, the author delivers a consistent model of cultural sustainability for Aboriginal tourism in British Columbia, as based on inductive research outcomes. However, the author is aware of the limitations that typically accompany qualitative research, such as a certain degree of subjectivity, linguistic ambiguity, and a lack of representativity (Atieno, 2009).

Since the author is not of Canadian nationality and, therefore, does not belong to the descendants of former generations of Europeans that took land away from the Aboriginal Peoples of Canada during settlement, nor is the author an Aboriginal Canadian, a certain level of neutrality as an outsider was accredited from the informants towards the author.

Table 1. Criteria and indicators for cultural sustainability (author's illustration).

Criterion	Indicator
Good self-governance	(a) A self-governance model exists in the Aboriginal community (b) This self-governance model is in line with the Canadian law, e.g. via a treaty
Human rights	(a) Human rights are guaranteed by national law for all citizens (b) The implementation of further agreements (e.g. a treaty) incorporates the specific needs of the Aboriginal Peoples of Canada
Identity	(a) A clear concept of the identity of the Aboriginal Peoples of Canada exists (b) This identity concept is underpinned by control of the land
Control of the land	(a) The Aboriginal Peoples of Canada dispose of land (b) The ownership of land by the Aboriginal Peoples of Canada is settled in a treaty
Control of the tourism product	(a) The tourism product provides themes related to the Aboriginal Peoples of Canada theme (b) The tourism product is controlled by the Aboriginal Peoples of Canada
Market-ready tourism product	(a) The business education among the Aboriginal Peoples of Canada enables the development and management of market-ready tourism products (b) The Aboriginal Peoples of Canada can access capital and other resources to enable the management of market-ready tourism products
Authenticity	(a) Aboriginal communities decide on marketable cultural assets (b) An umbrella organization ensures that only community-approved tourism products are on the market

Results

As a research outcome, the following model of cultural sustainability for Aboriginal tourism in British Columbia was developed.

Good self-governance, human rights, identity, control of land and the tourism product, market-ready tourism products, and authenticity form the key criteria of cultural sustainability for tourism in British Columbia.

Good self-governance

Traditional forms of governance among the Aboriginal Peoples comprise a council of elders and a hereditary chief; these forms continue to exist in parallel to officially elected bodies (e.g. an elected chief, leading to dual governing structures) (Hinch & Butler, 2007). Furthermore, there are clan systems and family groups that hold influence which, together, form diverse forms of government among the 198 Aboriginal communities in British Columbia (interview with Informant C). The Centre for First Nations Governance developed a toolkit for effective self-governance consisting of the following five pillars: The People, The Land, Laws and Jurisdiction, Institutions, and Resources (Centre First Nations Governance, 2017b). Good leadership and the existence of marketable tourism assets are key success factors in Aboriginal tourism (interview with Informants A, F). Therefore, the following two indicators are suggested when establishing the criteria for good self-governance:

- A self-governance model exists in the Aboriginal community; and
- this self-governance model is in line with the Canadian law (e.g. via a treaty).

Human rights

Human rights are a prerequisite for cultural sustainability. In general, they are guaranteed by Canadian law, but to specify and respect the needs of the Aboriginal Peoples of Canada (e.g. as defined in the Larrakia Declaration) (WINTA, 2017), a treaty provides even more clarity (PATA, 2015).

Therefore, the following two indicators are suggested to establish the human rights criterion:

- Human rights are guaranteed by national law for all citizens; and
- the implementation of further agreements (e.g. a treaty) must incorporate the specific needs of the Aboriginal Peoples of Canada.

Identity

The image of the Aboriginal Peoples of Canada still bears the legacy of an imposed, fictive identity of 'Indianness' based on economic, cultural, and sociopolitical inequalities that persist as stereotypes – sometimes with a racist connotation, sometimes with romantic attributes like 'Earth's original ecologists' (Nicholson, 1997). Many Aboriginal groups follow spiritual principals, and they embody the concept of Mother Earth and the traditional wisdom of the elders (Williams & O'Neil, 2007). The strong connection with the land, representing the ancestors' territory, is important for Aboriginal identity, too:

> The deep connection to the Land is vital to First Nations. The relationship between the People and the Land is an integral component of First Nations governance. The authority and identity of the Nations come from and are tied to the Land. It is the Land that gives us our deep sense of place and our sense of self. (Centre First Nations Governance, 2017b)

An authentic tourism product (cf. 5.6) is based on a firm Aboriginal identity and allows those in the industry to make appropriate decisions regarding what to select as a tourism product without damaging cultural integrity. Moreover, tourism – if managed in the right way – has the potential to overcome stereotypes and strengthen Aboriginal identity, leading to a renaissance of language, art, and other cultural aspects due to the level of respect and interest the tourist has towards Aboriginal culture. This will generate a deepened sense of pride toward the culture in Aboriginal communities (interviews with Informants C, I, J; Nepal, 2004; Notzke, 2006; Omand, 2015).

Therefore, the following two indicators are suggested for the Identity criterion:

- A clear concept of the existing identity of the Aboriginal Peoples of Canada; and
- this identity concept is underpinned by control of the land.

Control

Control of the land

The legislation in Canada helped the Aboriginal Peoples of Canada regain control of their land over the years (Hinch & Butler, 2009). Many Aboriginal Peoples in British Columbia live on land that is reserved for them (known as reserves), but which belongs to the Crown. These reserves are not necessarily their original territory, since displacement took place. Furthermore, the land of their ancestors holds great spiritual significance for the Aboriginal Peoples of Canada. In the pre-contact phase, the Aboriginal Peoples of Canada had property structures for the land, which was organized into sovereign nations (Centre First Nations Governance, 2017a). Thus, a complicated and interminable treaty negotiation process, the British Columbia Treaty Process, between the government and Aboriginal Peoples in British Columbia is currently underway to overcome the uncertainty regarding the use, tenure, management, and ownership of land in the Aboriginal community (INAC, 2016 and interview with Informant A). There are positive examples of co-management and partnership with the government, including that with the Nuu-chah-nulth First Nations on Vancouver Island or the Haida Gwaii Nations. The latter successfully negotiated conditions regarding visitor management and benefits for the community (Nepal, 2004). Overall, 65 Aboriginal Peoples in British Columbia participated or currently participate in the treaty process, which is comprised of six stages; four Nations reached stage six (i.e. a signed treaty), containing – amongst other agreements – access to land. This is where land title and self-governance were put into effect. These four First Nations are the Maa-nulth First Nations, the Tla'amin Nation, the Tsawwassen First Nation, and the Yale First Nation (BC Treaty Commission, 2017).

Thus, control of land (and water; Notzke, 1994) is not only a prerequisite for setting up a tourism business (Notzke, 1994; Williams & O'Neil, 2007), but it is highly interlinked with one's cultural background and identity (Anderson, Dana, & Dana, 2006). Furthermore, the Mother Earth concept is aligned with the idea of ecological sustainability, also known as traditional environmental knowledge; this concept has always been inherent to the Aboriginal mindset (interviews with Informants A, B, C, D, E, H, I, J; Deutsche Gesellschaft für Vereinte Nationen e. V., 2017; Notzke, 1994).

Therefore, the following two indicators are suggested for the control of land criterion:

- Aboriginal Peoples have land; and
- the ownership of land among the Aboriginal Peoples is settled in a treaty.

Control of the tourism product

The model of Hinch & Butler (Hinch & Butler, 2007) defines a 'culture controlled' form of tourism as a type of tourism that comprises an Indigenous theme under Indigenous control. When an Indigenous theme is absent, it still can be 'diversified Indigenous'; in cases of low Indigenous control, the authors speak of 'culture dispossessed' tourism. When pertaining to cultural sustainability, the latter concept is the least favourable: Aboriginal culture is displayed in tourism without the influence of the actual originators, which leaves the door wide open to cultural expropriation and infringement of Indigenous intellectual property rights and copyright (Hinch & Butler, 2007). Furthermore, asymmetrical relations between hosts and tourists, also referred to as neo-Colonial, may lead to uncontrolled cultural commodification that essentially prostitutes cultural integrity (Dorsey et al., 2004; Jamal & Stronza, 2008). A code of visitor conduct is a measure that can counteract such developments (Nicholson, 1997; Notzke, 2006). Museums displaying Aboriginal culture provide some ambiguity related to this issue; on the one hand, historical artefacts are turned into tourism products and are commodified – sometimes in museums in other countries beyond the control of the Aboriginal Peoples of Canada. Conversely, artefacts are preserved and explained to non-Aboriginal People. One perspective that is even more conflict rests in the different perspectives related to cemeteries: to scientists, cemeteries serve as archaeological sites; however, to the Aboriginal Peoples of Canada, cemeteries are sacred grounds of the ancestors (protocol participant observation Museum of Vancouver, MOV, 17 September 15). Sometimes, pressure arises to show something new to tourists; thus, ceremonies or artefacts that were designated as protected, and which needed to remain secret were in danger of being exposed (interview with Informant D). 'New Age Travelers,' in particular, may seek Aboriginal spirituality and act in an intrusive manner (Notzke, 2006).

It is important that the Aboriginal Peoples do not only offer single businesses along the tourism value chain; rather, they should aim to gain control over the entire value chain of an Aboriginal tourism product. The Aboriginal Art Galleries in Gastown and the Museum of Anthropology at the University of British Columbia (MOA at UBC), both in Vancouver, are examples where white Canadians still have the final say in the process; while the art in the galleries is produced by the Aboriginal Peoples of Canada, the salespersons and shop owners are white Canadians, as are the guides who interpret Aboriginal culture in the MOA at UBC (participant observation protocols 11 September 2015 and 31 August 2015). Examples where the Aboriginal Peoples have greater control of the Aboriginal tourism product throughout the value chain include the Squamish Lil'wat Cultural Center in Whistler, the Cedar Root Gallery, Musqueam Cultural Center, the Skwachàys Lodge, the restaurant Salmon n'Bannock (all four in Vancouver), the Spirit Ridge at Nk'Mip Resort (Osoyoos), and the St. Eugene Golf Resort and Casino (Cranbrook). However, some of these businesses (like the Cedar Root Gallery or the Musqueam Cultural Center) are located far away from typical tourist spots (participant observation protocols 15 September 2015, 23 September 2015, 25 September 2015, 26 September 2015, 27 September 2015, and 29 September 2015).

Therefore, the following two indicators are suggested for the control of the tourism product criterion:

- The tourism product provides themes related to the Aboriginal Peoples of Canada; and
- the tourism product is controlled by the Aboriginal Peoples of Canada.

Market-ready tourism product

Aboriginal tourism products need to be market ready to survive in the long run (interview with Informant C; AtBC, 2017; DBC, 2014). This economic pillar of cultural sustainability in Aboriginal tourism is hindered by limited human resources, limited marketing, and limited capital (Williams & O'Neil, 2007).

Therefore, the following two indicators are suggested for the market-ready tourism product criterion:

- Business education among Aboriginal Peoples of Canada enables the development and management of market-ready tourism products; and
- the Aboriginal Peoples of Canada have access to capital and other resources that enable the management of market-ready tourism products.

Authenticity

Authenticity in Aboriginal tourism can be assured by authenticity protocols; here, the community decides what is displayed for the tourist and what is protected (and therefore hidden) from visitors.

The AtBC decided not to market Aboriginal tourism products that are not approved by the community. Thus, authenticity in Aboriginal tourism is simply defined by the respective Aboriginal community and may include all types of mixtures of contemporary or traditional cultural elements (interview with Informant C; AtBC, 2017).

Therefore, the following two indicators are suggested for the authenticity criterion:

- Aboriginal communities decide on marketable cultural assets; and
- an umbrella organization ensures that only community-approved tourism products are on the market.

To summarize the results, the following indicators can be assigned to the model's criteria in Table 1.

Discussion

The criteria detailed above, as well as their associated indicators, intend to serve as a framework through which to develop and assess cultural sustainability in tourism in British Columbia. This framework can be modified or enhanced by stakeholders, destination management organisations, or single tourism businesses, based on their specific needs. The indicators of the respective criteria are gradually progressing. The first indicator is always a first step towards fulfilling the respective criterion, and the second indicator is either an enhancement (e.g. criterion control of land) or complementary (e.g. criterion control of the tourism product). Of course, the criteria are interdependent: control of the land is a requirement for both controlling the tourism product and identity. In addition, authenticity forms part of a market-ready tourism product and identity as well. Similarly, good self-governance may promote a market-ready tourism product, ensure human rights, and advocate for control of the land.

Although the situation for Aboriginal Peoples differs from country to country or, in Canada, from province to province, this framework (which was developed for British Columbia) can be transferred to other Aboriginal contexts, for instance, such as in those cases where the human rights criterion may be of higher importance than in British Columbia (PATA, 2015). Being a niche in terms of offer and demand, Aboriginal tourism is hardly in danger of developing towards a tourism mass market, preserving Aboriginal tourism from further commodification by mass tourism and thus conserving a unique cultural tourism product.

Conclusion

The framework of cultural sustainability for Aboriginal tourism in British Columbia that was developed in this paper may help to facilitate the management and preservation of Aboriginal culture when it intersects with tourism. The criteria and indicators of the framework intend to develop and maintain a form of tourism that is culturally sustainable. To verify this empirically may lead to further research and further evolution of the framework.

Cultural sustainability is a prerequisite for the identity and authenticity of the Aboriginal Peoples and for the tourism product itself. The state of Aboriginal tourism in British Columbia is much better than the situation of other Aboriginal communities offering tourism initiatives around the world. If cultural sustainability is maintained, tourism has the potential to develop as a powerful tool to further enhance and encourage the economic development and political empowerment of the Aboriginal Peoples, which would be in line with the latest announcement on a new legal framework for Indigenous Peoples by the Canadian Prime Minister, Justin Trudeau (CBC Politics, 2018).

Acknowledgement

I would like to thank two anonymous reviewers for their comments.

Disclosure statement

No potential conflict of interest was reported by the authors.

Funding

This work was supported by HTWG Konstanz.

References

Anderson, J. M. (1991). Problems with tourism development in Canada's eastern Arctic. *Tourism Management, 12*, 209–220.

Anderson, R. B., Dana, L. P., & Dana, T. E. (2006). Indigenous land rights, entrepreneurship, and economic development in Canada: 'Opting-in' to the global economy. *Journal of World Business, 41*, 45–55.

AtBC. (2017). *Aboriginal tourism action plan 2017–2018*. Vancouver.

Atieno, O. P. (2009). An analysis of strengths and limitation of qualitative and quantitative research paradigms. *Problems of Education in the 21st Century, 13*, 13–18.

ATS. (2018, February 2). Retrieved from https://www.aboriginaltravelservices.com

Axelsson, R., Angelstam, P., Degerman, E., Teitelbaum, S., Andersson, K., Elbakidze, M., & Drotz, M. K. (2013). Social and cultural sustainability: Criteria, indicators, verifier variables for measurement and maps for visualization to support planning. *AMBIO, 42*(2), 215–228.

BC Treaty Commission. (2017, November 16). Retrieved from http://www.bctreaty.ca/negotiation-update

Beckerman, Z., & Kopelowitz, E. (2008). *Cultural education and cultural sustainability: Minority, diaspora, indigenous and ethno-religious groups in multicultural societies*. London: Routledge, Education, (Book abstract).

Beddoe, R., Constanza, R., Farley, J., Garza, E., Kent, J., Kubiszewski, I., ... Woodward, J. (2009). Overcoming systemic roadblocks to sustainability: The evolutionary redesign of worldviews, institutions, and technologies. *Proceedings of the National Academy of Sciences, 106*, 2483–2489.

Besserab, D., & Ng'andu, B. (2010). Yarning about yarning as a legitimate method in indigenous research. *International Journal of Critical Indigenous Studies, 3*(1), 37–50.

Blundell, V. (1993). Aboriginal empowerment and souvenir trade in Canada. *Annals of Tourism Research, 20*, 64–87.

Boyd, S. (2001). Cultural heritage tourism in Canada: Opportinities, principles and challenges. *Tourism and Hospitality Research, 3*(1), 211–233.

Burton, R., & Paragahawewa, U. (2011). Creating culturally sustainable agrienvironmental schemes. *Journal of Rural Studies, 27*(1), 95–104.

Butler, R., & Hinch, T. (2007a). Conclusions: Key themes and issues in indigenous tourism. In R. Butler & T. Hinch (Eds.), *Tourism and indigenous peoples: Issues and implications* (pp. 319–331). Amsterdam: Butterworth-Heinemann.

Butler, R., & Hinch, T. (Eds.). (2007b). *Tourism and indigenous peoples: Issues and implications*. Amsterdam: Butterworth-Heinemann.

Canada's forced schooling of aboriginal children was 'cultural genocide', report finds. (2015). *New York Times*.

CBC Politics. (2018, February 21). Retrieved from https://www.facebook.com/CBCPolitics/videos/1970022789693213

Centre First Nations Governance. (2017a, November 17). Retrieved from http://fngovernance.org/timeline

Centre First Nations Governance. (2017b, November 22). Retrieved from http://fngovernance.org//publication_docs/Five_Pillars_CFNG.pdf

Choy, Y. K. (2004). Sustainable development and the social and cultural impact of a dam-induced development strategy – the Bakun experience. *Pacific Affairs, 7*(1), 50.

Cole, S. (2005). Action ethnography: Using participant observation. In B. W. Ritchie, P. Burns, & C. Palmer (Eds.), *Tourism research methods: Integrating theory with practice* (pp. 63–73). Wallingford: CABI.

Culture 21. (2011). *Lobbying for culture as the 4th pillar of sustainable development in the process of the Rio+20 summit*. Agenda 21 for culture. United Cities and Local Governments – Committee on culture.

Destination British Columbia. (2014). *Aboriginal cultural tourism – sector profile*. Vancouver.

Deutsche Gesellschaft für Vereinte Nationen e. V. (2017, November 28). Retrieved from http://nachhaltig-entwickeln.dgvn.de/meldung/indigene-voelker-vorbilder-fuer-eine-nachhaltige-entwicklung/

Dorsey, E. R., Steeves, H. L., & Porras, L. E. (2004). Advertising ecotourism on the internet: Commodifying environment and culture. *New Media and Society*, *6*(6), 753–779.
Dyer, P., Aberdeen, L., & Schuler, S. (2003). Tourism impacts on an Australian Indigenous community: A Djabugay case study. *Tourism Management*, *24*, 83–95.
Elbakidze, M., Angelstam, P. K., Sandström, C., & Axelsson, R. (2010). Multi-stakeholder collaboration in Russian and Swedish model forest initiatives: Adaptive governance toward sustainable forest management? *Ecology and Society*, *15*(2), 14.
Empacher, C., & Wehlung, P. (2002). *Soziale Dimensionen der Nachhaltigkeit*. Frankfurt am Main: ISOE-Studientexte.
Farsani, N. T., Coelho, C., & Costa, C. (2011). Geotourism and geoparks as gateways to socio-cultural sustainability in Qeshm Rural Areas, Iran. *Journal of Tourism Research*, *17*(1), 30–48.
Fennell, D. A. (2008). Ecotourism and the myth of indigenous stewardship. *Journal of Sustainable Tourism*, *16*(2), 129–149.
Global Sustainable Tourism Council. (2016, November 22). *Global sustainable tourism criteria*. Retrieved from http://www.gstcouncil.org/en/gstc-criteria-hotels-tour-operators-destinations/sustainable-tourism-gstc-criteria.html
Godelier, M. (2017). Die Anthropologie. Aus der Seele, die es nicht gibt, einen Menschen machen, der sie übertrifft. *Lettre*, *1*, 102–117.
Golafshani, N. (2003). Understanding reliability and validity in qualitative research. *The Qualitative Report*, *8*(4), 597–607.
Hawkes, J. (2001). *The fourth pillar of sustainability: Culture's essential role in public planning*. Melbourne: Common Ground Publishing Pty Ltd and Cultural Development Network Victoria.
Hinch, T., & Butler, R. (2007). Introduction: Revisiting common ground. In R. Butler & T. Hinch (Eds.), *Tourism and indigenous peoples: Issues and implications* (pp. 1–12). Amsterdam: Butterworth-Heinemann.
Hinch, T., & Butler, R. (2009). Indigenous tourism. *Tourism Analysis*, *14*(1), 15–27.
Indigenous and Northern Affairs (INAC), Government of Canada. (2016, December 23). Retrieved from http://www.aadnc-aandc.gc.ca/eng/1100100032288/1100100032289
Indigenous and Northern Affairs (INAC), Government of Canada. (2017, November 28). Retrieved from https://www.aadnc-aandc.gc.ca/eng/1100100013785/1304467449155
Jamal, T., & Stronza, A. (2008). 'Dwelling' with ecotourism in the Peruvian Amazon: Cultural relationships in local-global spaces. *Tourist Studies*, *8*(3), 313–335.
Johnston, A. (2000). Indigenous peoples and ecotourism: Bringing indigenous knowledge and rights into the sustainability equation. *Tourism Recreation Research*, *25*(2), 89–96.
Johnston, A. M. (2006). *Is the sacred for sale?* London, VA: Earthscan.
Kahakalau, K. (2004). Indigenous heuristic action research: Bridging western and indigenous research methodologies. *Multidisciplinary Research on Hawaiian Well-Being*, *1*(1), 19.
Kong, L. (2009). Making sustainable creative and cultural space in Shanghai and Singapore. *Geographical Review*, *99*(1), 1–22.
Kutzner, D., Wright, P. A., & Stark, A. (2009). Identifying tourists' preferences for Aboriginal tourism product features: Implications for a northern First Nation in British Columbia. *Journal of Ecotourism*, *8*(2), 99–114.
The Larrakia Declaration on the Development of Indigenous Tourism. (2012). (2017, November 5). Retrieved from http://www.winta.org/wp-content/uploads/2012/08/The-Larrakia-Declaration.pdf
Lemelin, R. H., & Blangy, S. (2009). Introduction to the special issue on Aboriginal ecotourism. *Journal of Ecotourism*, *8*(2), 77–81.
Li, Y. (2000). Ethnic tourism – a Canadian experience. *Annals of Tourism Research*, *27*(1), 115–131.
Louis, R. P. (2007). *Can you hear us now? Voices from the margin: Using indigenous methodologies in geographic research* (Aboriginal Policy Research Consortium International (APRCi). Paper 175).
Lynch, M.-F., Duinker, P., Sheeehan, L., & Chute, J. (2010). Sustainable Mi'kmaw cultural tourism development in Nova Scotia, Canada: Examining cultural tourist and Mi'kmaw perspectives. *Journal of Sustainable Tourism*, *18*(4), 539–556.
Mason, K. (2004). Sound and meaning in Aboriginal tourism. *Annals of Tourism Research*, *31*(4), 837–854.
Mccartney, G., & Osti, L. (2007). From cultural events to sport events: A case study of cultural authenticity in the dragon boat races. *Journal of Sport Tourism*, *12*(1), 25–40.
More Than Green – the multimedia encyclopedia of environmental, social, economic and cultural sustainability in the urban environment. (2016, November 22). Retrieved from http://www.morethangreen.es/en/ideology/four-sustainabilities-cultural-economic-social-environmental-sustainability/#sthash.oBEw9za5.dpbs
Murphy, K. (2012). The soical pillar of sustainable development: A literature review and framework for policy analysis. *Sustainability: Science, Practice and Policy*, *8*(12), 15–29.
Nassauer, J. I. (2004). Monitoring the success of metropolitan wetland restorations: Cultural sustainability and ecological function. *Wetlands*, *24*(4), 756–765.
Nepal, S. K. (2004). Indigenous ecotourism in Central British Columbia: The potential for building capacity in the Tl'azt'en nations territories. *Journal of Ecotourism*, *3*(2), 173–194.

Nepal, S. K. (2005). Limits to indigenous ecotourism: An exploratory analysis from the Tl'azt'en territories, Northern British Columbia. In C. Ryan & M. Aicken (Eds.), *Indigenous tourism* (pp. 111–126). London: Elsevier.
Nicholson, H. N. (1997). Collusion, collision or challenge? Indigenous tourism and cultural experience in British Columbia, Canada. In P. E. Murphy (Ed.), *Quality management in urban tourism* (pp. 115–136). New York, NY: John Wiley & Sons.
Notzke, C. (1994). *Aboriginal peoples and natural resources in Canada*. North York: Captus Press.
Notzke, C. (1999). Indigenous tourism development in the arctic. *Annals of Tourism Research, 26*(1), 55–76.
Notzke, C. (2004). Indigenous tourism development in southern Alberta, Canada: Tentative engagement. *Journal of Sustainable Tourism, 12*(1), 29–54.
Notzke, C. (2006). *'The stranger, the native and the land': Perspectives on indigenous tourism*. Concord: Captus Press.
Oh, D.-S., Kim, K.-B., & Jeong, S.-Y. (2005). Eco-industrial park design: A Daedeok Technovalley case study. *Habitat International, 29*, 269–284.
Omand, G. (2015, September 16). *First nations tourism is growing*. Vancouver Sun, C4.
O'Neil, B., Williams, P., Morten, K., Kuhln, R., Gan, L., & Payer, B. (2015). *National Aboriginal tourism research 2015: Economic impact of Aboriginal tourism in Canada*. Vancouver: Aboriginal Tourism Association Canada (ATAC).
Pacific Asia Travel Organisation. (2015). *Indigenous tourism and human rights in Asia and the Pacific region*. Bangkok: Author.
Ryan, C., & Aicken, M. (Eds.). (2005). *Indigenous tourism: The commodification and management of culture*. Oxford: Elsevier.
Schmiechen, J. (2006). *Indigenous tourism research agenda*. Darwin: Charles Darwin Univsersity.
Seale, G. E. (1996). A perspective from Canada on heritage and tourism. *Annals of Tourism Research, 23*(2), 484–488.
Soini, K., & Birkeland, I. (2014). Exploring the scientific discourse on cultural sustainability. *Geoforum; Journal of Physical, Human, and Regional Geosciences, 51*, 213–223.
Stewart, E. J., & Draper, D. (2009). Reporting back research findings: A case study of community-based tourism research in northern Canada. *Journal of Ecotourism, 8*(2), 128–143.
Throsby, D. (2003). Cultural sustainability. In R. Towse (Ed.), *A handbook of cultural economics* (pp. 183–186). Cheltenham: Edward Elgar.
Trimarchi, M. (2004). Regulation, integration and sustainability in the cultural sector. *International Journal of Heritage Studies, 19*(5), 410–415.
Tuhiwai Smith, L. (2005). On tricky grounds: Researching the native in the age of uncertainty. In N. K. Denzin & Y. S. Lincoln (Eds.), *The handbook of qualitative research* (pp. 85–108). Thousand Oaks, CA: Sage.
United Nations Declaration of the Rights of Indigenous Peoples. (2007). (2017, November 12). Retrieved from http://www.un.org/esa/socdev/unpfii/documents/DRIPS_en.pdf
UN: United Nations Permanent Forum on Indigenous Issues. (2016, February 6). *Factsheet who are indigenous peoples?* Retrieved from http://www.un.org/esa/socdev/unpfii/documents/5session_factsheet1.pdf
Van Mansvelt, J. D. (1997). An interdisciplinary approach to integrate a range of agrolandscape values as proposed by representatives of various disciplines. *Agriculture, Ecosystems and Environment, 63*(2–3), 233–250.
Vileniske, I. G. (2008). Influence of built heritage on sustainable development of landscape. *Landscape Research, 33*(4), 425–437.
Von der Pahlen, M. C., & Grinspoon, E. (2002). Promoting traditional uses of medicinal plants as efforts to achieve cultural and ecological sustainability. *Journal of Sustainable Forestry, 15*(1), 81–93.
Wall, A. (2008). Reinventing the wheel? Designing an Aboriginal recreation and community development program. *Canadian Journal of Native Education, 31*(2), 70–93.
Weaver, D. (2010). Indigenous tourism stages and their implications for sustainability. *Journal of Sustainable Tourism, 18*(1), 43–60.
Weaver, D. B. (2016). Beyond convention: Reimagining indigenous tourism. *Tourism Geographies, 18*(1), 28–37.
Whitford, M., & Ruhanen, L. (2016). Indigenous tourism research, past and present: Where to from here? *Journal of Sustainable Tourism, 24*(8–9), 1080–1099.
Williams, P. (2013). *AtBC Blueprint Strategy: Tourism performance audit report*. Vancouver: Aboriginal Tourism Association of British Columbia.
Williams, P., & O'Neil, B. (2007). Building a triangulated research foundation for indigenous tourism. In R. Butler & T. Hinch (Eds.), *Tourism and indigenous peoples: Issues and implications* (pp. 40–57). Amsterdam: Butterworth-Heinemann.
Williams, P., & Richter, C. (2002). Developing and supporting European tour operator distribution channels for Canadian Aboriginal tourism development. *Journal of Travel Research, 40*, 404–415.
Williams, P., & Stewart, J. K. (1997). Aboriginal tourism development: Assessing latent demand from France. *The Journal of Tourism Studies, 8*(1), 25–41.
Wilson, S. (2001). What is indigenous research methodology? *Canadian Journal of Native Education, 25*(2), 175–179.
WINTA. (2017, November 22). Retrieved from http://www.winta.org/wp-content/uploads/2012/08/The-Larrakia-Declaration.pdf
Zeppel, H. (2006). *Indigenous ecotourism: Sustainable development and management*. Wallingford: CABI.

List of interviews
Informant A. (2015, September 4). Academia, male, interview by the author.
Informant B. (2015, September 5). Academia, Aboriginal Canadian, male, interview by the author.
Informant C. (2015, September 8). Tourism industry, Aboriginal Canadian, male, interview by the author.
Informant D. (2015, September 8). Academia, male, interview by the author.
Informant E. (2015, September 21). Academia, female, interview by the author.
Informant F. (2015, September 24). Tourism industry, female, interview by the author.
Informant H. (2015, September 27). Tourism industry, female, interview by the author.
Informant I. (2015, September 29). Tourism industry, Aboriginal Canadian, female, interview by the author.
Informant J. (2015, September 9). Academia, female, interview by the author.

List of participant observation protocols
Art Galleries in Gastown. (2015 September 9). Vancouver, observation by the author.
Cedar Root Gallery. (2015, September 15). Vancouver, observation by the author.
Ktunaxa Interpretive Center. (2015, September 25). Cranbrook, observation by the author.
Lil'wat Cultural Center. (2015, September 27). Whistler, observation by the author.
MOA at UBC. (2015, August 31). Vancouver, observation by the author.
Museum of Vancouver, MOV. (2015, September 17). Vancouver, observation by the author.
Musqueam Cultural Center. (2015, September 15). Vancouver, observation by the author.
Nk'Mip Cultural Desert Center. (2015, September 23). Osoyoos, observation by the author.
Royal British Columbia Museum. (2015, September 10). Victoria, observation by the author.
Salmon n'Bannock. (2015, September 29). Vancouver, observation by the author.
Skwachàys Lodge. (2015, September 26). Vancouver, observation by the author.

Developing indigenous tourism in the *bomas*: critiquing issues from within the Maasai community in Tanzania

Kokel Melubo and Anna Carr

ABSTRACT

With safari attractions dominating the African tourism sector, many tourism studies in Africa have focused on wildlife and wilderness resourced-based tourism, with a secondary body of literature exploring cultural aspects of tourism. Recently many countries in the region, including Tanzania, have started to recognize the potential for stand-alone, culture-based tourism to diversify the tourism industry and for rural, economic development purposes. Drawing upon insights from local stakeholders, primarily members of the Maasai community this study identifies the critical challenges for providing indigenous cultural tourism in the Ngorongoro Conservation Area, Tanzania. Qualitative data for the study were collected through interviews and site visits to eight Maasai *bomas* (cultural tourism villages or settlements). Thematic analysis of the interviews revealed significant limiting factors affecting the development of indigenous cultural tourism amongst Maasai in Tanzania. The findings indicate that achieving success in indigenous cultural tourism is therefore challenging, and needs to incorporate specific community-based strategies which can facilitate the development of Maasai cultural tourism.

Introduction

Indigenous tourism is one of the fastest expanding segments of tourism across the world (Fletcher, Pforr, & Brueckner, 2016; Saarinen & Rogerson, 2015; Whitford & Ruhanen, 2016). Indigenous culture-based tourism can be a powerful activity providing community development and infrastructure opportunities, employment and economic benefits (Saarinen & Rogerson, 2015; Whitford & Ruhanen, 2016). Cultural tourism is also praised for its potential to enable communities to continue living in remote areas whilst participating in the tourism economy without moving away from ancestral lands, thus maintaining their place in culturally significant landscapes (Carr, 2007; Colton & Whitney-Squire, 2010; Fletcher et al., 2016). International research has focused on various aspects affecting the supply and demand factors of indigenous and cultural tourism. Within the context of national parks and protected areas, researchers have explored the various implications of nature conservation on indigenous communities including the pros and cons of engaging in related tourism activities (Eagles & McCool, 2002; Stevens, 1997; Zeppel, 2007, 2009). In their examination of past and future research around indigenous tourism, Whitford and Ruhanen (2016, p. 1091) observed that

> It is imperative that academic researchers gain a more comprehensive understanding of indigenous tourism from the perspective of indigenous stakeholders because the complexity of indigenous tourism should ideally

be approached in an iterative, adaptive and flexible style and affected stakeholders need to be part of the research process, knowledge creation and outcomes.

This study, led by a Maasai researcher amongst members of his extended community network, seeks such understanding by exploring the perspectives of his own people – the Maasai community – as tourism stakeholders. Previous studies have explored the impact of tourism development on the Maasai from political, marketing, anthropological and geographical perspectives, for instance by analyzing the performative nature or imagery of Maasai tourism experiences (Akama, 2002; Bruner & Kirshenblatt-Gimblett, 1994). No previous studies have examined the perspectives of the Maasai people themselves, where they reveal the constraints they face when engaging in indigenous cultural tourism as a means of diversifying their traditional pastoral livelihoods. This paper seeks to counter the limited understanding of Maasai perspectives of being in the business of tourism; importantly delving into the dynamics of how business activities are influenced by their cultural values or challenged by what seems to be ongoing potential for displacement of Maasai communities from protected areas.

Background

In Tanzania, culture-based tourism is currently advertised by Tourism Tanzania alongside the more dominant marketing of national park and wildlife-based tourism. The natural wonders of Mt Kilimanjaro, the Serengeti Plains and Ngorongoro Conservation Area (NCA) are prominent in online marketing and 'Cultural Tourism' is a category noted as potentially benefiting the social and economic wellbeing of its citizens in rural settings (Anderson, 2015). The inclusion of culture-based tourism came after a realization that tourists are searching for broader experiences that are not focused on wildlife resources (Anderson, 2015). A formal government policy for cultural tourism development began in 1995 in Tanzania, with support from the Dutch aid agency Stichting Nederlandse Vrijwilligers (SNV) (Salazar, 2012). This encouraged local communities to develop economically viable tourism activities such as campsites, home-stays, traditional foods, and visits to cultural and historical sites. Thus, today, promoting culture-based tourism is one of the core roles of the Tanzania Tourist Board (TTB). In particular, the TTB has established a unit responsible for the development and promotion of culture-based tourism into fully fledged tourism products (Salazar, 2012; URT, 2010). Furthermore, other government tourism-related agencies, such as the Tanzania National Parks (TANAPA), have integrated culture-based tourism in their corporate strategic plans.

Prior to formal government support to establish culture led tourism, the Maasai community in the NCA had started culture-based tourism activities, taking advantage of the abundance of tourists visiting this area. The primary activities included producing art, jewelry, and beaded handicrafts, as well as exhibitions of cultural activities including traditional dancing and costume, for tourists frequenting the Arusha-Serengeti tourist highway. Capitalizing on this opportunity, Maasai communities, supported by NCA management, developed cultural *bomas* (also known as *manyattas* in Kenya), which are tourism villages with the express purpose of conducting regular performances and tours as cultural tourism activities (Bruner, 2001).

The Maasai people are aware that their culture has been commodified as an economically valuable asset used to yield financial returns through tourism, with scholars critiquing instances of the tourism industry appropriating Maasai culture through staged performances or the exoticization of the people (Bruner, 2001; Mgonja, Sirima, & Mkumbo, 2015; Salazar, 2006, 2012). Studies by Okello and Yerian (2009) and Salazar (2012) indicated that experiencing the Maasai is one of the primary motivations for visitors to Tanzania and Kenya. In particular, Okello and Yerian's (2009) examination of visitor satisfaction in Northern Tanzania found that tourists come to visit the Serengeti-Ngorongoro ecosystem primarily to experience the 'big five wildlife', but 'among those who considered culture as an important tourist magnet, about 83% were attracted mainly by the Maasai compared with other indigenous cultures of the area' (Okello & Yerian, 2009, p. 615).

The Maasai are traditionally a pastoral peoples living across southern Kenya and northern Tanzania (McCabe, 2003). The Maasai people speak the Eastern Nilotic Maa tonal language with different dialects. Although the Maa-speaking groups such as *Ilarusa* and *Irparakuyo* practice crop farming, keeping livestock remains a pivotal livelihood strategy for many Maasai (McCabe, 2003). Maasai depend on livestock products including milk, meat and blood for their main nutritional intake. Census records of the Maasai population in Tanzania are unavailable (as ethnicity is not recorded in the Tanzanian national census), but it is estimated that Maasai in Tanzania number approximately 500,000 people (Maasai Association, 2018).

Together with other local ethnic groups such as the Akiye, Hadzabe and Datoga, Maasai have identified and described themselves as members of the national indigenous communities' movement (Hodgson, 2011; Salazar, 2015). Through the network of indigenous communities, the Maasai are able to address critical challenges facing them such as loss of land tenure, forced settlement, under-representation in politics, poor social services and cultural disparagement (Goldman, 2011; Hodgson, 2011). Similar to other indigenous communities, Maasai feel they have experienced discrimination and been subject to displacement and the loss of land and natural resources on which cultural identity depends (Goldman, 2011; Igoe, 2006). This claim has, however, been challenged by some observers including government officials noting that identifying as 'indigenous' has ethnic connotations which might be used to strengthen political opposition to the status quo (Hodgson, 2011; Igoe, 2006). Goldman noted a 'participation gap' for Maasai who have been isolated by conservation efforts and excluded from political decision-making, thus becoming 'Strangers in their own land' (Goldman, 2011, pp. 71–72).

In response to the Government disapproval of the political implications of identifying as 'indigenous', the Maasai have resolved to replace the phrase 'indigenous' with 'pastoralists' and a focus on 'pastoralists' rights and 'pastoralists' livelihoods (Hodgson, 2011; Igoe, 2006). Such a repositioning has enabled the Maasai community to pay attention to local, national and international socio-economic and political affairs and commentaries such as access to political self-determination, social services, water, employment and education. Although visitor demand to experience Maasai culture has substantially increased over the years within the Ngorongoro Area limited academic attention has investigated the success (or otherwise) of tourism in the communities (Buzinde, Kalavar, & Melubo, 2014; Salazar, 2012) and this paper addresses this gap in knowledge.

Tourism in Maasai communities in Tanzania

There have been both positive and negative impacts of tourism on the Maasai for instance through the economic benefits arising from employment as tour guides (Melubo & Buzinde, 2016). Similarly, the interaction between Maasai communities and tourists purchasing handicrafts has positively elevated the status of women as craftspeople (Buzinde et al., 2014; Melubo & Lovelock, 2017). Through tourism microenterprises, particularly the sale of handicrafts, a significant number of Maasai women have been empowered economically, challenging the patriarchal culture where men were the sole breadwinners of families, often restricting women's business initiatives as observed by Dutt, Grabe, and Castro (2016). Buzinde et al. also observed patriarchal attitudes such as 'women used to be under men's feet depending on them for food and other basic needs … with money from tourism they can buy animals and food' (2014, p. 30). In Lollondo Area, Ngorongoro, the Maasai community uses the revenues obtained from tourism for the construction of schools, healthcare centers, village offices, and for salaries to pay medical staff or village game scouts who patrol the areas against poachers (Snyder & Sulle, 2011). Similarly, the Maasai in the Ngorongoro area use a significant amount of revenue from tourism to support schools, infrastructure and bursaries for students to study within and outside the NCA (Charnley, 2005; OleNasha, 2014). Furthermore, the Maasai residents through the Ngorongoro Pastoral Council receive approximately 2–4% annually of the 100 plus billion Tanzanian Shillings (USD 40–50 million) collected from tourism in the NCA (Melubo & Lovelock, in press). Nevertheless, because of poor technical skills in tourism, only a small proportion of Maasai people are employed full time at the NCA and lodges surrounding the Ngorongoro area apart from instances where, based on the perception of Maasai as brave, honest and fierce people, some Maasai are hired as *walinzi* (security guards).

Negatively, tourism and conservation have been associated with the loss of prime pasture land (Bruner, 2001; McCabe, 2003). Because of the flourishing wildlife and iconic touristscapes and sites, Maasai inhabitants of their cultural landscapes have encountered increased competition for space with conservationists, hunting outfitters and tour operators (Buzinde et al., 2014; Charnley, 2005; Gardner, 2016; Okello & Yerian, 2009; Snyder & Sulle, 2011). Thus 'conservation' has negative connotations for the Maasai people as they are displaced from their traditional lands and continue to be concerned that future land loss may arise (Goldman, 2011; OleNasha, 2014). Surprisingly, 'tourists have been granted access and investors are allowed to build lodges and other facilities catering for tourists' (Sirima & Backman, 2013, p. 721). Snyder and Sulle (2011) report that the Maasai community in Northern Tanzania have lost trust for anything labeled 'conservation,' even if promoted as socially and environmentally responsible, because such terms have been used to deprive their valuable pasture and water resources.

One case in point is the Loliondo division in Ngorongoro in which the community has lost significant land to conservation and hunting investors (Gardner, 2016). In Esilalei and Oltukai villages in Monduli district, Buzinde et al. (2014) reported that investors, often foreigners, initially go to the villages as tourists and later, supported by the government, they gain status as investors who push the communities away. Gardner (2016) and Sirima and Backman (2013) note that because of tourism and conservation the Maasai people bear the majority of the burden, in that they are being marginalized through the loss of access to natural assets (land and water). With limited livelihood opportunities, the Ngorongoro Maasai community is arguably the poorest in Northern Tanzania on all economic development indices including severe food insecurity (Gardner, 2016). Consequently, as highlighted by Galvin, Boone, McCabe, Magennis, and Beeton (2015), the Maasai in the Ngorongoro often experience incidents of malnutrition – poverty within the Maasai community largely an outcome of ongoing land rights issues.

Tourism has, theoretically, provided a space for the socio-economic empowerment of the Maasai. In early tourism interactions, handicrafts and the commodification of traditional objects, manifested in the sale of leather, beaded and wooden engraved items such as necklaces, bracelets, belts and swords. Bruner and Kirshenblatt-Gimblett (1994) discussed the cultural politics of Maasai participating in dance performances stereotyping them as 'noble savages' thus bringing to life imagery of historical 'colonialism.' Bruner (2001) critiqued the presentation of Maasai culture for tourism interests, for instance, the objectification of Maasai men performing to tourists in 'Out of Africa' staged cocktail parties that perpetuated stereotypes (Bruner, 2001, p. 3). During the 1990s traditional dances were regularly performed every time a tourist vehicle entered into Maasai settlements (Akama, 2002). Recently researchers have noted negative sociocultural impacts indicating consequences for cultural moral values, for instance, the creep of unwanted behaviors such as drug-taking, smoking and recorded incidents of commercial sex were observed amongst Maasai in a study by Buzinde et al. (2014). In their paper, the Maasai community at Esilalei village, Monduli district, bemoaned that through tourism some young herders have developed the habit of waiting on roads to be photographed in exchange for candy or money (Buzinde et al., 2014). Thus, while the Maasai community has had benefits from tourism, a common finding in various studies is that tourism and conservation activities have also brought sociological, economic and psychological problems for the Maasai (Akama, 2002; Bruner, 2001; Buzinde et al., 2014; Charnley, 2005; Gardner, 2016; Goldman, 2011; Snyder & Sulle, 2011).

Sadly tourism, particularly nature-based tourism with associated conservation management approaches such as the relocation of Maasai from traditional lands, has regularly failed to consistently fulfill the needs of the Maasai community; instead contributing to marginalization and ongoing human rights violations (Akama, 2002; Buzinde et al., 2014; Gardner, 2016; Goldman, 2011). Notwithstanding the published evidence of such negative impacts upon the Maasai community in

Northern Tanzania, the Tanzania Tourist Board and other agencies continue to promote tourism as a tool for development and poverty reduction. Careful management is needed to ensure benefits arise as despite two decades of increased Maasai involvement in cultural tourism there is limited evidence that tourism initiatives address the equity and poverty challenges facing Maasai.

Maasai *bomas*

The establishment of *bomas* (Figure 1) gives tourists an opportunity to encounter and experience the Maasai culture without impacting upon the inhabitants. Tourists have the opportunity to view traditional dancing, to observe aspects of the Maasai way of life and to purchase Maasai handcrafts such as traditional weapons, jewelry and artwork (NCAA, 2006). For the Maasai, 'pseudo' *bomas* are a strategy to capture the tourist dollar, and are an alternative source of income when the livestock sector performs poorly. Kipuri and Sørensen (2008) argue that cultural *bomas* have become a new economic sector and 'safe net' that prevent many Maasai people from descending into abject rural poverty. Alongside cultural *bomas*, Maasai people benefit as guides for walking safaris and campsite, allowing the Maasai people to earn 50% of the fees (the other half goes to the NCA) through employment and participating directly by supplying guiding services or renting donkeys to carry luggage.

Today there are 10 officially registered *bomas* along the Serengeti-Ngorongoro highway. Although there are variations on the shape and style of the *bomas*, most of them reproduce traditional Maasai homesteads (not the contemporary homes typically owned by many twentieth-century Maasai). These traditional buildings are dome-shaped, mud houses encircled by cattle *kraal* (enclosures).

The setting of the *bomas* is a useful one through which to explore Maasai perspectives of cultural tourism in Tanzania, presenting *bomas* community members' experiences and personal knowledge of how they have lived indigenous cultural tourism, thus furthering understandings of indigenous stakeholders' hopes and concerns about issues surrounding indigenous tourism. By focusing on the Maasai perspectives the study also contributes to knowledge on Maasai understandings of 'empowerment' which has been relatively under-explored in tourism studies.

Study area and methods

The study was carried out in seven cultural *bomas*, all of which are located along the highway that transports tourists from Arusha town to the Serengeti National Park and Ngorongoro Area. Permission to use the *bomas* registers, which list each individual members' name, gender and role (e.g. dancers, handicraft producers and local guides), was used to randomly select participants for interviews. A total of 104 people (83 males and 21 females) residing across seven cultural settlements were identified as relevant participants (Table 1). The selection of 10 to 15 people from each cultural

Figure 1. Cultural settlements (Lead Author).

Table 1. Interview participants.

Selected participants $n = 104$				
Ward	*Cultural settlement*	Male	Female	Total
Endulen	Ilailupa	13	3	16
	E-nasinya	11	4	15
Ngoile	Loonkooku	13	2	15
Olbalbal	Ndemwa	9	3	12
Nainokanoka	Irkeekpusi	13	2	15
Ngorongoro	Seneto	12	3	15
Misigiyo	Elerai	12	4	16
Total		83	21	104

settlement enabled the potential for both similarities between *bomas* but also a range of differing perspectives to be attained (rather than focusing on 1 or 2 *bomas*).

The majority of male respondents were youth or middle-aged males of the *Irkorianka* and *Nyangulo* age groups.[1] Both age groups are presently the most active singing and dancing groups in Tanzanian Maasai cultural settlements. For the Maasai, age is referenced alongside cultural traditions and practices such as circumcision. In adhering to the Maasai protocols and customs, and to give them freedom to speak out, women were interviewed separately from men (Buzinde et al., 2014) (Table 1). In addition to interviews with *bomas* residents, purposive sampling was carried out to interview key informants from the Pastoral Council. Four key informants (a tourism officer, a community development officer and two *diwanis*, (political elected leaders representing administrative wards) were interviewed. All interviews took place outdoors, often under a large shade tree, using a naturally conversational style, and lasted between 30 and 40 minutes.

The *bomas* were visited to gain insights into the range of tourism activities conducted at the *bomas*. During the field visits, photographs were taken of tourism infrastructure and sites to provide practical illustrations of what constitutes core tourism activities within the *bomas* and surrounds.

Thematic analysis (Braun & Clarke, 2006) was employed to analyze the data from the interviews. The transcripts of interviews were translated and transcribed into English from Maa or Swahili by the lead author. Themes that emerged during the interviews were identified and crosschecked and (De Crop, 2004; Denzin & Lincoln, 2005). Direct quotations from participants are used in this paper in their English translation to provide insights into the participants' experiences of tourism. The following section presents the Maasai voices by discussing the Maasai participants' perspectives and observations through their personal articulations of constraints affecting Maasai-centric cultural tourism initiatives.

Findings

From the interview transcripts eight themes emerged that affect the development and success of Maasai cultural tourism. The themes are discussed in the following order: political interference; shortage of staff; quality of products and services; marketing; under-representation of women in leadership; financial management; social services; and competition.

Political interference

Evidence of political interference was seen by nearly half of the research participants as hampering the development of Maasai-led cultural tourism. Decision-making on the management of the *bomas* was concentrated in the hands of elected *diwanis* (ward councilors) who use their political powers to decide who is to join the *bomas*, the number of *bomas*, and location of *bomas* with limited consultation with the villagers, nor consideration of the ecological fragility and sensitivity of the areas.

> Some diwanis have manipulated the leadership of the bomas to their favour. They ignore the guidelines by PC. They decide who is to chair the bomas and most of which are his political followers. They dictate the distribution mechanism and expenditures of the collections. They decide who gets what and when. (Participant 1, Ngorongoro Ward)

Often each cultural village is required to host 50 residents for a year, but in some instances bomas exceed this number and may contain 100 plus residents. In order to spread tourism revenues more equally, community tourism guidelines call for a rotation scheme which allows residents to stay for a year at the *bomas*, but some *diwanis* encouraged their friends to overstay.

Following the 2015 Tanzanian General Election a number of cultural settlements have increased from one to two in some wards.

In addition, contrary to the community tourism guidelines developed by the Ngorongoro Pastoral Council (NPC) in 2016 some *diwanis* appointed friends or relatives to lead the *bomas*, which could be seen as positive in terms of indigenous intergenerational well-being but controversial if western concepts of nepotism and corruption are considered. The guidelines empower the ward authorities to nominate members of the *bomas* based on the expression and willingness of the individual to join and also the wealth status (number of livestock) of members. Furthermore, some *diwanis* dictate the expenditure of the revenue collected, including the salaries of the people. Related to political interference is the question of leadership within the villages. Some village leaders are powerful and overshadow others, the former perceiving themselves as the ultimate decision-makers according to interview participants. Some leaders were accused of misusing money from the village coffers.

Shortage of staff

Over a third of the research participants identified an insufficient number of professional staff at the *bomas* to run the tourism businesses. It was argued that limited literacy and numeracy skills of the leaders, as well as other members, at the cultural villages were challenges. The cultural *bomas* policy requires leaders serving at each village to have acquired a basic education that enabled them to write, count and read. Other credentials, such as English proficiency and any other western language were also considered an assett for public relations officers of the villages. Contrary to the requisite qualification matching any village responsibilities, few leaders speak foreign languages let alone are equipped with marketing and leadership skills to govern the *bomas*. Noting this weakness, one respondent observed that 'After a very brief shallow introduction, the majority of cultural guides are rather sharp to invite tourists to walk around the village and take pictures.' For instance, local guides were sometimes observed as being unable to either explain the socio-structural organization of Maasai settlements or provide interpretive information about the Maasai around contemporary lifestyles, traditional architecture, dress, hairstyles, crafts, history, dance and food. This need for interpretation and customer care skills (when hosting foreign guests), was observed by one participant:

> To tell the truth, we learn on job. No formal training has been done. Some guides have local formal education but can hardly speak and communicate well in English but we keep trying. I think people need some training on business and managerial skills and how to design quality products. (Participant 4, Ngoile Ward)

With culture-based tourism businesses being regarded highly as a livelihood source lack of the requisite skills, education and capacity building for those managing cultural tourism enterprises could therefore affect the service quality delivered by the Maasai people. The development of professional or skilled staff in areas such as marketing, guiding and communication skills needs serious attention.

Quality of products and services

Maintaining standards and quality assurance of tourism products is becoming increasingly important for business sustainability globally, including in indigenous cultural tourism settings (Whitford

& Ruhanen, 2016). A quarter of the research participants considered poor design, quality and packaging of the activities, products and services provided at the *bomas* was undermining the development of the tourism – 'Tourists like buying quality Maasai handmade items but our people sometimes fail them; they fail to design and produce good ones'. Participants noted all *bomas* along the tourist highway had near-identical crafts, guided tours, beaded items, traditional dances and foods on offer, with limited unique features differentiating *bomas*. In addition, some of the items sold, such as handcrafts, were perceived by interview participants to be dusty, smoky, smelly and sooty – not attractive items for tourists to buy (though potentially this is the trade-off for authenticity). According to participants, bomas are money-making ventures and therefore many of the activities and products offered are to the taste of the tourists. They further noted that some *bomas* do not provide sufficient opportunity for tourists to meet and talk to Maasai people or even to experience performances which might increase visitor understanding. Interview participants felt there were opportunities for further product development so that tourists could experience Maasai warriors' fire-making or dancing and observe women making jewelry or handicrafts such as decorated gourds however this could also perpetuate stereotypes that limit understanding of Maasai culture.

Marketing

The lack of marketing strategies was considered by some participants as a constraint to achieving high quality, culture-centric tourism. Strategies to promote tourism products to targeted markets are essential (Keyser, 2009). Indeed, the NCA management used to prominently feature the Maasai-led cultural activities in their website and travel brochures, supporting small groups of local Maasai to attend national and international trade fairs and exhibitions. Maasai culture used to be a strong marketing image for Ngorongoro Tourism. Signposts were installed along the road to provide directions to villages but were removed so that the *bomas* would look less commercial. At present, the word of mouth and occasional visits by local leaders to foreign markets overseas has become a powerful marketing tool to communicate the value and appeal of Maasai tourism. There is currently no local level attempt to promote community-led tourism in the area. Research participants noted the current efforts by the NCA are fragmented with limited coordination amongst the NCA alongside communities, tour operators and guides:

> The NCA is doing less to promote aggressively cultural tourism. The NCA website which supposed to market Maasai practices features less of people but of wildlife ... we know, what that white people (tourists) admire Maasai; they come to Tanzania to see and enjoy our culture and wildlife. It is self-selling because little support from NCA to speak about our bomas, no brochures and website. (Participant 9, Ngorongoro Ward)

The NCA management was also observed to facilitate the participation of unskilled people in the promotion of traditional cultural activities at local and international trade fairs (e.g. Saba Saba and Nane Nane), rather than sending people who actually worked in the *bomas*. To recognize the potential of Maasai-led cultural tourism, ensuring informed participants from the communities is needed for these products and activities to remain competitive.

Under-representation of women in leadership

Female research participants expressed concern that Maasai women are underrepresented in the running and management of cultural *bomas* and some male participants shared this concern. In all seven cultural villages not one had a woman in a leadership position such as a chairperson, accountant or public relations officer. The local guiding and interpretation businesses which attract tips from tourists had no Maasai women on the teams. Yet, women make up over 60% of the residents at the *bomas*. In addition, Maasai women are the skillful crafters of almost all jewelry and handcrafts such as beadwork including necklaces and bracelets (whilst spears, clubs and knives are made by men). Women also enact an active role in welcoming tourists and with the performances of songs and dance. Most handicrafts, including beadwork, are sold to tourists at the *bomas*.

> The perception that women are weaker than men is strong within the (Maasai) community. Community gives less regard to women. In every bomas, women are the majority but no woman is a chairperson of a boma or as a public relation officer. (Participant 2, Misigiyo Ward)

Traditionally Maasai women lack power within business settings, a finding reflected in the work of Dutt et al. (2016). Maasai culture is patriarchal, privileging men as breadwinners and leaders, whilst women are expected to behave conservatively and responsibly, as mothers and housekeepers. Regular household chores for Maasai women include collecting fuelwood, fetching water, obtaining, processing, and cooking food for the households. Reasons advanced by interview participants as to the low female participation included that Maasai women still feel uncomfortable and shy when potentially assuming leadership positions. Men are perceived as not honoring women's' ideas as women traditionally have low levels of access to schooling, limited educational attainment and often cannot speak foreign languages compared to men. The reality is that many women are the principal teachers for many preschools at the *bomas* staged for tourists. The reasons for women working as teachers include, firstly, that tourists are tempted to donate money when they find 'mothers' teaching pupils. Secondly, mothers are perceived as having 'caring hearts' to support children. The low participation of women in managing cultural tourism initiatives reflects the dominant gender roles among the Maasai, consequently halving the opportunities to benefit communities through tourism initiatives. As such, while cultural tourism initiatives present advantages for women such as increased incomes, it is still the men who take the 'lion share' (ironically) of the leadership roles and revenue related to tourism.

Financial management

Lack of transparency in the collection of fees was also mentioned by interview respondents as a significant impediment to sustainable cultural tourism ventures amongst Maasai. There are three ways in which the *bomas* capture tourism dollars: donations to preschools, entrance fees and selling handicrafts. In terms of entrance fees, participants noted that although the official entrance fee might be US$20 per vehicle (or sometimes per tourist), tour drivers often demand commission reimbursement from villages of TZS10,000 (US$5). When villagers object to such arrangements, tour drivers simply take tourists to another village that would accept such deals.

> The disturbing thing is less that 20% of what tourists pay goes as entrance fee reaches into our coffers. About 70 to 80% of that entrance fee goes back to drive guides who convinced tourists to visit us. This is bad common practice and nobody is doing something to discourage it. (Participant 2, Misigiyo Ward)

As such, the tour drivers take advantage of the harsh environment and stiff competition amongst the broader Maasai community's *bomas*, as they know Maasai are dependent on customers for financial payments (as opposed to trading or bartering goods and services). Since receipts for payments are not issued it is very difficult to trace how much overall income is turned over at villages and economic leakage impacts the communities severely. In one case study *bomas* the preschool's donation box is managed by the school teacher or village chairperson and in some instances the *diwanis*. Participants complained about the lack of transparent reporting on revenue and expenditure.

Some villages encourage individuals to sell from their own stalls whilst in others it is local guides or public relations officers who sell the items on behalf of owners as 'middlemen'. Although the selling of artworks and handicrafts by non-owners reduces conflict or misunderstandings with customers, participants complained that middlemen (*the local guides*) tend to sell their valuable items in powerful foreign currency (Euros and dollars) but pay the owners in Tanzanian shillings and pocket the profit.

The intended benefit-sharing mechanism expected from earning tourist dollars was noted for being erratically managed, as observed by the following participant:

> We know how much money is generated because we too count tourist vehicles coming to this boma but we are not told how they money is distributed … No transparency on the use. (Participant 6, Ngorongoro Ward)

For example, the community tourism guidelines (NPC, 2016) direct that entrance fees are split between the *bomas* for the provision of services such as purchasing water. Similarly, fees are divided between salaries (60%), the respective village governments (35%), and the administering ward governments for community development projects (5%). However, the power to allocate the earned dollars within communities, including the village dwellers, lies in the hands of the *diwanis* as village leaders. Participants alleged that a lion share of the 60% for *bomas* is retained by this latter group, for example, at Elerai *boma*, leaders receive over TZS 80,000, and the rest of the members receive TZS 30,000. For some villagers, a substantial amount of revenue is directed towards the purchase of food (maize flour), water, medication and supporting education.

Social services

Galvin et al. (2015) observed that although NCA is mandated to care for conservation, tourism and people, from its inception the management has insufficiently addressed social services for the Maasai people. While improving social services of communities living inside or outside protected areas is globally emphasized, this is sadly lacking in the NCA. Three issues emerged regarding social services, all of which were negative. Firstly, while the Maasai people regard tourism as a means to ease negative socio-economic circumstances, funding the provision of social services such as health centers and water supplies remains a significant challenge across the Ngorongoro ecosystem. Over a third of the participants reiterated that the NCA residents are probably amongst the poorest population in Tanzania in the provision of social services:

> It is absurd that despite many rich tourists frequenting the Ngorongoro the Maasai living here are one of the poorest. Food insecurity and malnutrition is a big challenge here. (Participant 3, Ngorongoro Ward)

Research participants revealed that water is a scarce resource across the entire Ngorongoro ecosystem, especially in the dry season from August through November. Shortage of water resources cost *bomas* a substantial amount of total revenue collections. For example, in 2016, Elerai *boma* allocated TZS 3,200,000 (15% of their collection) to hire water tanks from private individuals. In addition, transportation to nearby health care centers such as Endulen Hospital and Karatu town was perceived to be highly costly:

> Yes, Ngorongoro is a World Heritage Site, yes there are many tourists visiting and bring much money but we have serious shortage of water, only two secondary school and one hospital in the whole of Tarafa ya Ngorongoro (Ngorongoro Division). (Participant 6, Ngoile Ward)

A source of power for cooking and lighting was mentioned as another concern. The Maasai rely on firewood and dry cattle dung for energy and cooking as they do not have access to electricity from the national grid. Most cultural *bomas* do not keep enough cattle to produce dung for plastering the *enkajijik (huts)* and for cooking fuel. In addition, the overreliance of trees for fencing the villages and firewood has reduced accessible vegetation and forest resources:

> In the past it was much easier to collect firewood and poles for building of huts; shrub was plentiful and much closer. People cut bushes for cooking and fencing bomas. (Participant 4, Ngorongoro Ward)

Participants, particularly women, reported having to walk up to 5 kilometers to locate firewood. Initiatives to build energy-saving stoves and solar energy that may help reduce indoor smoke and drudgery of women's daily work have yet to be realized. The general hygiene of the cultural villages was also noted as poor. Most villages have no constructed pit latrines, making *bomas* vulnerable to sanitary diseases and associated problems.

Finally, research participants indicated that none of the *bomas* have waste management procedures or facilities for the disposal of non-organic garbage such as cans, bags, plastic bottles, and other containers.

Competition

Participants noted that along the Arusha-Serengeti roadside and in some towns such as Mto wa Mbu and Karatu, there has been a proliferation of numerous clusters of curio shops selling, amongst other items, Maasai crafts. Many of these craft items are sold at exorbitant prices, but they are perceived to be relatively superior in quality in comparison to those sold at cultural villages. As noted by one respondent 'the lodges and other curio shops sell well made and polished vinyago compare to us (Maasai at the boma).' Similarly, some cultural enterprises have resolved to waive or lower the entrance fees to certain *bomas*, making them the most preferred by tour drivers. Furthermore, some curio shops offer commissions of around TZS10,000–20,000 to safari drivers who bring customers to them.

When combined with the fact that Maasai cultural tourism is marketed and experienced separately to Tanzania's nature-based wildlife and safari tourism products there is competition between cultural tourism and nature tourism products for the visitor dollar (Tanzania Tourist Board, 2018). Statistics (URT et al., 2016) indicate that about 98% of visitors to Ngorongoro Serengeti Ecosystem are motivated by wildlife viewing, including the historically most dangerous African animals known as the 'Big 5' (elephant, lion, buffalo, and rhino and leopard) (Okello & Yerian, 2009) and thus Maasai indigenous cultural tourism could be viewed as niche. Nevertheless, the Tanzania Tourist Board's website markets cultural tourism opportunities as beneficial for the Maasai with the aspiration that:

> Various local communities run their own cultural programmes and welcome visitors to their homes, bringing income directly to local community while giving local people an opportunity to showcase their way of life to the outside world. This creates mutual understanding and friendships between tourists and local people, offering tourists from all over the world the possibility to experience Tanzania's cultural diversity and providing local people in various rural areas the opportunity to build sustainable livelihoods. (Tanzania Tourist Board, 2018)

Improving the marketing competitiveness alone is however not sufficient to realize the 'sustainable livelihoods' of Maasai from cultural tourism when considered alongside the issues identified in the preceding themes.

Conclusion

This study revealed that whilst some benefits do arise from tourism activities the Maasai NCA residents have experienced being disempowered socially and economically. That the long-term sustainability of Maasai-led cultural tourism is impeded by challenges such as competing attractions and businesses (both between Maasai themselves and non-Maasai attractions or providers). There is a prevalent view that capacity building is required with low skill sets amongst staff responsible for managing initiatives. Other concerns expressed were that there is both poor quality of products and services and provision of social services which would ideally spread the benefits of cultural tourism. Local political interference and gender issues surrounding the poor representation of women were also raised by participants and reflect issues recognized in previous research examining Maasai (Anderson, 2015; Charnley, 2005; Dutt et al., 2016; Goldman, 2011; Salazar, 2015). Nonacademic studies, including those by organizations such as Tourism Concern have exposed similar critical challenges; however, apart from Dutt et al's study on Maasai women in business, this research is one of the few to present the perspectives and viewpoints of insiders based on observations and interviews with participants from within the Maasai *bomas*. Given that this study has focused mainly on the constraints of indigenous-led tourism from the perspective of Maasai people in Tanzania; future studies could evaluate the perceptions of other culturally rich NCA non-Maasai indigenous communities such as those of the Hadza and Datoga peoples. This study's findings suggest that although Maasai women are taking part in developing indigenous tourism, it appears that there are still barriers to women's empowerment and participation in tourism. Future research could also be conducted to explore ways by which Maasai women might be further empowered and meaningfully involved in social and economic realms.

The findings of this study support previous observations that tourism is an opportunity for Maasai communities to diversify their livelihoods and fostering sustainable community development. Maasai-tourist relationships require the mitigation of inappropriate visitor and tourism management behavior. This study suggests that realizing the benefits of tourism remains a thorny issue as previously noted by Salazar (2012) and Snyder and Sulle (2011). Evidently, the Maasai community members interviewed in this study do not believe they are receiving sufficient benefits from cultural tourism initiatives such as purposeful built *bomas*, yet the Maasai shoulder much of the cost of tourism and conservation through loss of their pastoral traditions when facing continual displacement from traditional grazing areas for their cattle.

In order for Maasai-led tourism to succeed in a manner that improves the welfare of the Maasai people, the study findings suggest the following.

First, there is a fundamental need for capacity building through community-centric initiatives so that local people could expand their knowledge and expertise of tourist interests and expectations. Communities could benefit from cooperative approaches to develop visitor strategies to guide sustainable tourism. To recognize the positive impacts of tourism amongst the Maasai communities, collaboration with the NCA management alongside local government by developing policies and strategies to market tourism products and services would be a preferable option. Tourism managers, workers and local guides, who in this context represent and speak for the community and their values (Salazar, 2012), need to be fully aware of community and tourist expectations. Similarly, there is a need for capacity building amongst employees whilst strategic selection of leaders should occur in a transparent manner to avoid perceptions or accusations of nepotism and gender inequality. This need for equity throws up challenges to Maasai ways of traditionally ordering society or doing business. Modernizing and developing community services whilst conserving the ecological balance of the natural environment that surrounds communities and villages through planning and implementation of conservation measures, could augment access to natural resources vital for Maasai daily life. The reality, however, is that although recognizing traditional forms of political leadership is a critical component in the sustainability of culture-centric tourism, it appears from this study that those community members with political powers within Maasai communities need to review how they traditionally allocate resources or responsibilities and share the socio-economic benefits of tourism. Tourism is thus simultaneously challenging and relying upon Maasai cultural traditions.

Note

1. An Age group is a corporate group of male who are initiated at one time through circumcision (Morton, 1979). The Irkorianka entered into an age set in 1996 and retired in 2015. Nyangulo are the newly forming age-set.

Disclosure statement

No potential conflict of interest was reported by the authors.

References

Akama, J. (2002). The creation of the Maasai image and tourism development in Kenya. Cultural tourism in Africa: Strategies for the new millennium. In *Proceedings of the ATLAS Africa international conference* (pp. 43–54). Kenya: Mombasa .

Anderson, W. (2015). Cultural tourism and poverty alleviation in rural Kilimanjaro, Tanzania. *Journal of Tourism and Cultural Change, 13*(3), 208–224.

Braun, V., & Clarke, V. (2006). Using thematic analysis in psychology. *Qualitative Research in Psychology, 3*(2), 77–101.

Bruner, E. M. (2001). The Maasai and the lion king: Authenticity, nationalism, and globalization in African tourism. *American Ethnologist, 28*(4), 881–908.

Bruner, E. M., & Kirshenblatt-Gimblett, B. (1994). Maasai on the lawn: Tourist realism in East Africa. *Cultural Anthropology, 9*, 435–470.

Buzinde, C. N., Kalavar, J. M., & Melubo, K. (2014). Tourism and community well-being: The case of the Maasai in Tanzania. *Annals of Tourism Research, 44*, 20–35.

Carr, A. M. (2007). Strengthening of identity through participation in tourism guiding. In R. Butler & T. Hinch (Eds.), *Tourism and indigenous peoples* (2nd ed., pp. 113–127). Oxford: Butterworth-Heinemann.

Charnley, S. (2005). From nature tourism to ecotourism? The case of the Ngorongoro conservation area, Tanzania. *Human Organization, 64*(1), 75–88.

Colton, J. W., & Whitney-Squire, K. (2010). Exploring the relationship between aboriginal tourism and community development. *Leisure/Loisir, 34*(4), 261–278.

De Crop, A. (2004). Trustworthiness in qualitative tourism research. In J. Phillimore & L. Goodson (Eds.), *Qualitative research in tourism* (pp. 156–169). London: Routledge.

Denzin, N., & Lincoln, Y. (2005). *The sage handbook of qualitative research* (3rd ed.). Thousand Oaks, CA: Sage.

Dutt, A., Grabe, S., & Castro, M. (2016). Exploring links between women's business ownership and empowerment among Maasai women in Tanzania. *Analyses of Social Issues and Public Policy, 16*(1), 363–386.

Eagles, P. F., & McCool, S. F. (2002). Tourism, protected areas and local communities. In P. F. Eagles & S. F. McCool (Eds.), *Tourism in national parks and protected areas: Planning and management* (pp. 187–210). Oxon: CABI.

Fletcher, C., Pforr, C., & Brueckner, M. (2016). Factors influencing indigenous engagement in tourism development: An international perspective. *Journal of Sustainable Tourism, 24*(8–9), 1100–1120.

Galvin, K. A., Boone, R. B., McCabe, J. T., Magennis, A. L., & Beeton, T. A. (2015). Transitions in the Ngorongoro conservation area: The story of land use, human well-being, and conservation. In A. R. E. Sinclair, K. L. Metzger, S. A. R. Mduma, & J. M. Fryxell (Eds.), *Serengeti IV: Sustaining biodiversity in a coupled human-natural system* (pp. 483–512). Chicago: University of Chicago Press.

Gardner, B. (2016). *Selling the Serengeti: The cultural politics of safari tourism*. Athens: University of Georgia Press.

Goldman, M. J. (2011). Strangers in their Own land: Maasai and wildlife conservation in northern Tanzania. *Conservation and Society, 9*(1), 65–79.

Hodgson, D. L. (2011). *Being Maasai, becoming indigenous: Postcolonial politics in a neoliberal world*. Bloomington: Indiana University Press.

Igoe, J. (2006). Becoming indigenous peoples: Difference, inequality, and the globalization of east African identity politics. *African Affairs, 105*(420), 399–420.

Keyser, H. (2009). *Developing tourism in South Africa*. Cape Town: Oxford University Press.

Kipuri, N., & Sørensen, C. (2008). *Poverty, pastoralism and policy in Ngorongoro: Lessons learned from the Ereto I Ngorongoro pastoralist project with implications for pastoral development and the policy debate*. London: ERETO/IIED.

Maasai Association. (2018). Retrieved from http://www.maasai-association.org/maasai.html

McCabe, J. T. (2003). Sustainability and livelihood diversification among the Maasai of northern Tanzania. *Human Organization, 62*(2), 100–111.

Melubo, K., & Buzinde, C. N. (2016). An exploration of tourism related labour conditions: The case of tour guides in Tanzania. *Anatolia, 27*(4), 505–514.

Melubo, K., & Lovelock, B. (2017). Reframing corporate social responsibility from the Tanzanian tourism industry: The vision of foreign and local tourism companies. *Tourism Planning & Development, 15*(6), 672–691. doi:10.1080/21568316.2017.1395357

Melubo, K., & Lovelock, B. (in press). Living inside a UNESCO world heritage site: The perspective of the Maasai community in Ngorongoro conservation area, Tanzania. *Journal of Tourism Planning and Development.*

Mgonja, J. T., Sirima, A., & Mkumbo, P. J. (2015). A review of ecotourism in Tanzania: Magnitude, challenges, and prospects for sustainability. *Journal of Ecotourism, 14*(2–3), 264–277.

Morton, R. F. (1979). The structure of east African age-set systems. *Pula: Botswana Journal of African Studies, 1*(2), 77–102.

NCAA. (2006). *Ngorongoro conservation area general management plan (2006–2016)*. Arusha, Tanzania: Ngorongoro Conservation Area Authority, 171.

Ngorongoro Pastoral Council. (2016). Mwongozo wa Usimamizi na Uendeshaji wa Maboma ya Kitamaduni katika Hifadhi ya Ngorongoro, NCA.

Okello, M. M., & Yerian, S. (2009). Tourist satisfaction in relation to attractions and implications for conservation in the protected areas of the northern circuit, Tanzania. *Journal of Sustainable Tourism*, *17*(5), 605–625.

Olenasha, W. (2014). A world heritage site in the Ngorongoro conservation area: Whose world? Whose heritage? In S. Disko & H. Tugendhat (Eds.), *World heritage sites and indigenous peoples' rights* (pp. 189–220). Copenhagen: IWGIA.

Saarinen, J., & Rogerson, C. M. (2015). Setting cultural tourism in Southern Africa. *Nordic Journal of African Studies*, *24*(3&4), 207–220.

Salazar, N. B. (2006). *Touristifying Tanzania*: Local guides, global discourse. *Annals of Tourism Research*, *33*(3), 833–852.

Salazar, N. B. (2012). Community-based cultural tourism: Issues, threats and opportunities. *Journal of Sustainable Tourism*, *20*(1), 9–22.

Salazar, N. B. (2015). The Maasai as paradoxical icons of tourism (im)mobility. In A. C. Bunten, J. Chio, & N. H. H. Graburn (Eds.), *Cultural tourism movements: New articulations of indigenous identity* (pp. 56–72). Toronto: Toronto University Press.

Sirima, A., & Backman, K. F. (2013). Communities' displacement from national park and tourism development in the Usangu plains, Tanzania. *Current Issues in Tourism*, *16*(7–8), 719–735.

Snyder, K. A., & Sulle, E. B. (2011). Tourism in Maasai communities: A chance to improve livelihoods? *Journal of Sustainable Tourism*, *19*(8), 935–951.

Stevens, S. (Ed.). (1997). *Conservation through cultural survival: Indigenous peoples and protected areas.* Washington, DC: Island Press.

Tanzania Tourist Board. (2018). *Cultural tourism.* Retrieved from http://tanzaniatourism.go.tz/things-to-do/cultural-tourism

URT Ministry of Natural Resources and Tourism. (2010). *Guidelines for cultural tourism in Tanzania.* Dar es Salaam, Tanzania: United Republic of Tanzania, Ministry of Natural Resources and Tourism.

URT Ministry of Natural Resources and Tourism, Bank of Tanzania, Zanzibar Commission for Tourism, National Bureau of Statistics and Immigration Department. (2016). *The 2016 international visitors' exit survey report.* Dar es Salaam. Retrieved from http://www.nbs.go.tz/nbs/takwimu/trade/The2016_International%20_Visitors'_Exit_Survey_Report.pdf

Whitford, M., & Ruhanen, L. (2016). Indigenous tourism research, past and present: Where to from here? *Journal of Sustainable Tourism*, *24*(8–9), 1080–1099.

Zeppel, H. (2007). Indigenous ecotourism: Conservation and resource rights. In J. E. S. Higham (Ed.), *Critical issues in ecotourism: Understanding a complex tourism phenomenon* (pp. 308–348). Oxford: Butterworth-Heinemann.

Zeppel, H. (2009). National parks as cultural landscapes: Indigenous peoples, conservation and tourism. In W. Frost & M. Hall (Eds.), *Tourism and national parks: International perspectives on development, histories and change* (pp. 259–281). Oxon: Routledge.

Settler colonialism, Indigenous cultures, and the promotional landscape of tourism in Ontario, Canada's 'near North'

Bryan S. R. Grimwood, Meghan L. Muldoon and Zachary M. Stevens

ABSTRACT

Settler colonialism relies upon a logic of elimination that strives to dissolve Indigenous culture and title to ancestral lands. In Canada, settler colonialism has steered not only oppressive state policy directions, but also settler narratives that essentialize and displace Indigenous Peoples and cultural connections to land. Tourism is an especially potent social force through which such settler stories can be perpetuated and resisted. The purpose of this paper is to analyze how Indigenous culture is portrayed in the tourism promotional landscape of Ontario's 'near north', a rural leisure landscape for nature enthusiasts and second-home owners. While Anishinaabeg and Haudenosaunee First Nations have inhabited this area for generations, their cultural presence is often marginalized in development discourses, both historical and contemporary. Our study draws upon critical discourse analysis of websites produced by stakeholders associated with three case study sites – a Provincial Park, an annual cranberry Festival, and a major casino operation – to understand both the limits and opportunities of tourism in relation to maintaining and revitalizing Indigenous culture. In so doing, our paper contributes to critiques of settler colonial power relations and how these infiltrate tourism, and identifies pathways for disrupting the erasure of Indigenous cultures in tourism.

Introduction

The promotion of tourism is hardly benign. As a form of representation, tourism promotions operate within dynamic networks of values, meanings, power relations, histories, and ideologies, often serving to reinforce prevailing social structures and desires (Urry, 1992). This observation echoes in work showing how tourism imagery essentializes places and identities in ways generating mass appeal (e.g. Amoamo & Thompson, 2010; Cohen, 1993; Morgan & Pritchard, 1998), and related efforts on how visitors, residents, and the wider tourism complex interact in the production of destination imaginaries for particular forms of consumption (Aitchison, MacLeod, & Shaw, 2000; Pritchard & Morgan, 2000). All too often, when Indigenous Peoples and cultures are the subject of tourism promotions, the 'industry of difference, par excellence' (Hollinshead, 1998, p. 49) leans especially heavy on its capacity to homogenize and historicize. It is common to see, for instance, Indigenous lifeways being 'fossilized,' represented not in their actual form as dynamic cultural practices (Lee, 2017), but rather as artefacts well-suited for museums or other cultural storehouses where the customary tourist gaze is one locked on some romanticized version of difference (Hollinshead, 1992; Seiver & Matthews, 2016; Yang, 2011). Indeed, tourism promotions tend to code Indigenous Peoples as the remaining vestiges of a 'dying breed' (Yang, 2011, pp. 562–563) whose cultural dress, ceremonies, and customs can be preserved and celebrated through tourism. Indigenous cultures are, more often than not, positioned at the peripheries of modernity; their contemporary presence and agency made invisible (Braun, 2002).

Awhile back, Hollinshead (1992) argued that 'technologies of subjugation which are at work must not remain unrecognized' (p. 45). He challenged scholars to consistently expose oppressive discourses and how these serve to silence alternative ways of relating to tourism landscapes. In this paper, we take up this recurrent task by critically examining online promotions associated with three case study sites in Ontario, Canada, for imprints of settler colonialism and its discursive limits. As we elaborate below, settler colonialism is an ongoing project predicated upon the erasure of Indigenous Peoples from their homelands (Veracini, 2015). Despite advancements in literature on tourism and Indigenous Peoples (see e.g. Whitford & Ruhanen, 2016), there is in our view a dearth of tourism research that makes the articulation, disruption, or refusal of settler colonial power relations its explicit mandate [important exceptions include Amoamo and Thompson (2010), Cooke (2016; 2017), and Lee (2017)]. This paper helps fill this gap. Specifically, our purpose is to critically analyze how Indigenous cultures are portrayed in the tourism promotional landscape of Ontario's 'near north,' a regional and perceptually demarcated leisure landscape for nature enthusiasts and second-home owners. Anishinaabeg and Haudenosaunee First Nations[1] have inhabited, used, and moved through this area for generations. Their presence and cultures, however, are often marginalized in development discourses, which instead favor settler affinities and desires. Our study draws upon critical discourse analysis of online promotional materials produced by tourism brokers and Indigenous stakeholders associated with a provincial park, an annual cranberry festival, and a major casino operation. We aim to expose settler colonial power at work within this Canadian tourism scene and shed light on both the limits and opportunities of tourism in relation to maintaining and revitalizing Indigenous cultures.

Settler colonialism and tourism

Settler colonialism is a particular form of colonial domination whereby invasion transpires as 'a structure not an event' (Wolfe, 2006, p. 388). Settler colonial invasion, in other words, is not a historical moment of imperial conquest of Indigenous lands, but rather an ongoing form of occupation exerted through the enduring social, political, and economic structures built by the invading society (Veracini, 2015). Under settler colonialism, these structures, which include cultural norms and practices that are institutionalized through laws (e.g. private property) and social discourse (e.g. ideals of progress), advance a 'logic of elimination' (Wolfe, 2006, p. 387); one that perpetually seeks to dissolve Indigenous societies and title to lands in order to allow for the construction and continuation of a colonial society on the expropriated land base (see also Coulthard, 2014; Lowman & Barker, 2015). This requires settler societies to symbolically and literally erase Indigenous Peoples from the landscape. It also demands that settler societies suppress their moment of origin, working continuously to normalize and justify the existence of their settler nation-state over top of Indigenous societies (Cooke, 2016; Veracini, 2011, 2015).

In Canada, settler colonialism is advanced in overt and elusive ways. Canadian legislation and policy directions focused on the assimilation of Indigenous Peoples are, arguably, the most profound forms of state-sanctioned violence. The *Indian Act* of 1876, for instance, was crafted with the intention of dispossessing Indigenous Peoples of their land while also compelling them to renounce their 'Indian' status and become 'enfranchised' as full members of Canadian society (Coulthard, 2014). Amendments to this act in 1920, which made residential school attendance mandatory for Indigenous children, further fueled a process that government agents at the time declared would 'get rid of the Indian problem' (Regan, 2010, p. 238) and which, just recently, was officially recognized as contributing to cultural genocide (TRC, 2015). Stories that underpin Settler identity in Canada also contribute to Indigenous erasure (Lowman & Barker, 2015). The most exemplary is the one that begins with European explorers 'discovering' pristine lands – that is, *terra nullius* – and working hard to overcome exceptional struggles and establish a coherent and progressive society. This story denies the original presence of Indigenous Peoples and the crucial role they have played in Settler survival. It also bleeds into ongoing imaginaries of, and desires for, Canadian wilderness (Lowman & Barker, 2015). Importantly, while such erasures are consequential in very tangible ways for Indigenous Peoples, their effects are reliant upon, and strengthened through, their invisibility among Settlers (Cooke, 2016, 2017; Veracini, 2011). What Rifkin (2013) has described as 'settler common sense' is embedded within the mentality of everyday life, continually reinforcing its own normalization. Processes of power and domination, reinforced through narrative and the rule of law, allow Settlers to understand as 'given' their continued presence on and occupation of Indigenous lands (Rifkin, 2013). As Settlers, we feel entitled to the land we occupy, secure and sovereign. This is coupled by our tendency to disavow any history – let alone present-day responsibilities – as colonizers (Lowman & Barker, 2015). Consequently, Indigenous ways of knowing and being become 'buried' and settler values, knowledges, and cultural forms become discursively 'naturalized' within the national identity that grows out of them, rendering them at once pervasive and invisible (Cooke, 2016).

No doubt tourism is often complicit in marshaling settler colonial power, producing leisure destinations and pleasure peripheries in ways that support singular forms of consumption and experience, and imposing Settler desires and narratives on places and peoples. Research has shown, for example, how tourism disciplines visitors to perceive and experience Indigeneity within specific spatial ('untouched' nature) and temporal (historic) registers, which operate to obscure and deem out-of-place contemporary Indigenous practices, meanings, and mobilities within a landscape (Cooke, 2017; Erickson, 2018; Grimwood, Yudina, Muldoon, & Qiu, 2015; Hall & Tucker, 2004). That tourism development tends to be controlled by non-Indigenous Peoples and dominated by colonial systems and institutions are notable roots to the problem (Butler & Hinch, 2007). This translates, as Amoamo and Thompson (2010) observe, into tourism representations that stereotype particular characteristics of Indigenous cultures and communities with images catering to 'exotic' and 'primitive' 'otherness.' Indeed, tourism promotions and other marketing strategies are persuasive expressions of power that mediate how tourists see and engage with the world (Morgan & Pritchard, 1998).

Yet research also suggests that tourism holds promise as a means for enabling resistance to settler colonialism. The Indigenous tourism literature includes examples of how tourism can service Indigenous community-control and autonomy over lands (Holmes, Grimwood, King, & the Lutsel K'e Dene First Nation, 2016), cultural narratives and adaptations (Notzke, 1999; Ryan & Aicken, 2005), knowledge and value transmission (Higgins-Desbiolles, 2003; Walker & Moscardo, 2016), and economic and social development (Bennett, Lemelin, Johnston, & Koster, 2010; Seiver & Matthews, 2016). This literature also signals the demise of damaging development narratives that cast Indigenous communities as passive 'hosts' or 'locals' with homogenous experiences of being targeted by tourism as threat or opportunity (Viken & Müller, 2017). Specific to promotions, exemplary studies by Seiver and Matthews (2016) and Amoamo and Thompson (2010) illustrate how Indigenous tourism businesses and stakeholders are taking control of representations to present Indigenous destinations and identities in complex, multiple, and hybridized ways. These 'counter-representations' illustrate Indigenous Peoples' self-determined use of tourism to disrupt settler stereotypes and classifications of the Other (Amoamo & Thompson, 2010; Seiver & Matthews, 2016). If tourism is to realize its potential as a social force, including its capacity to foster peace and respectful cross-cultural understandings (Higgins-Desbiolles, 2003), we must be diligent in making space for Indigenous representations of culture, identity, and place-based relations. Part of this involves being consistently alert to, naming, and pushing back against the ways settler colonial power gets re-inscribed through tourism.

Study contexts

Critical investigations of settler colonialism have gained traction in Canada due to recent public dialogue and state-sponsored actions concerning reconciliation between Indigenous and Settler Peoples.

Much momentum flows directly from the 2015 report of the Truth and Reconciliation Commission (TRC, 2015), which was tasked by the federal government to learn from survivor and witness accounts of the Canadian Indian Residential School system in order to chart a pathway for healing and truth-telling in relationships between Settler Canadians and Indigenous Peoples. The residential schools existed for over 100 years – the last one closing in 1996 – and were deliberately designed to indoctrinate Indigenous children into the Euro-Christian Canadian mainstream by separating them from their families and cultures.[2] After six years and over 6,000 interviews, the Commission issued 94 Calls for Action aimed at moving Canadian society towards greater understanding and reconciliation with Indigenous Peoples. A central feature of these recommendations – which our current Liberal government has indicated its commitment to implementing – is to reaffirm the nation-to-nation relationships the underpin original agreements made between Indigenous Peoples and the Crown. This would require repudiating concepts and narratives (e.g. *terra nullius*) used to justify Settler occupancy of and sovereignty over Indigenous lands, and building and maintaining respectful relations between Indigenous and Settler Peoples (TRC, 2015). By taking a critical look at Canadian-based tourism promotions, a central aim of our research has been to understand how tourism is positioned to facilitate or impede these aspirations.

Research reported herein is part of a multifaceted study examining relationships between tourism and Indigenous livelihoods in the near north of the province of Ontario. While Ontario's near north may consist of certain state-sponsored districts or official destination marketing regions [e.g. our study focuses on sites within Ontario Tourism Regions 7 and 12], we use it herein to signify a perceptually demarcated leisure landscape for nature enthusiasts and second-home owners (Harrison, 2013). The near north is prominent in the imaginaries and desires of urban residents, especially those from Toronto, other southwestern Ontario cities, and American states across the border. Visitors of varying types – cottagers, hunters, ecotourists, youth summer campers and their families, lodge guests, snowmobilers – are drawn to the near north to experience particular constructions of nature, often coded as 'pristine' or 'wilderness' and quintessential to Canadian identity (Jasen, 1995). This tourism place-making has been, and continues to be, inscribed upon the territories of diverse Indigenous Peoples. Specifically, the near north is situated on the homelands of Anishinaabeg (meaning the 'Original Peoples') – which includes various Nations that share the native language *Anishinaabemowin* – and Haudenosaunee (meaning 'People of the Longhouse') – which includes various Nations that share one of the several Iroquoian languages (Gehl, 2013). In culturally distinctive ways, these Nations farmed, hunted, fished, trapped, traveled, and traded upon their expansive homelands for generations (Ray, 2016). These connections to land were fragmented as Canada implemented the *Indian Act's* reserve system to confine Indigenous territories to a fraction of their original size.

Three cases selected for this paper represent different modes of contemporary tourism punctuated by different scales of nature and different Indigenous communities. Case one considers *Algonquin Provincial Park*. Situated on the ancestral territory of Algonquin First Nations (Anishinaabeg), Algonquin Park is Ontario's emblematic macro-scale nature tourism destination. Established in 1893 and covering more than 7,600 square kilometers, Algonquin Park operates as a multi-zoned protected area with high-traffic tourist corridors in some areas, intensive logging and scientific research in others, and backcountry canoeing and hiking yet in others. In addition to its widespread use and popularity, Algonquin Park is an intriguing case because of current land claim contexts. After two and half decades of negotiations, the Algonquins of Ontario, which includes members of the Pikwàkanagàn First Nation located east of the park, have reached an agreement-in-principle with federal and provincial governments over rights to lands that were never extinguished (Ontario, 2018). While much of Algonquin Park lies within the land claim area, parties have agreed that no part of it will be included in the lands transferred back to Algonquin Peoples. It appears that the park's integrity as a protected area and nature-based destination is important for all to maintain.

Case two focuses on the annual *Cranberry Festival* in the town of Bala. Since 1984, this volunteer-driven event has attracted upwards of 20,000 visitors to the 'Cranberry Capital of Ontario' in the

District of Muskoka the weekend after Canadian Thanksgiving (mid-October). It has proven effective at extending visitation to the area at a time when the pace of summer and early autumn tourism has all but subsided. Festival spirit centers on harvesting, distributing, and experiencing micro-scale nature in the form of the cranberry, and is thus contingent upon agricultural economies. One of the two local farm suppliers over the years, Iroquois Cranberry Growers, was established in 1968 and owned and operated by the Wahta Mohawk First Nation (Haudenosaunee). During its prime, the Iroquois Cranberry Growers was the most successful Indigenous community-owned business in Canada.[3] Along with the other Bala-area farm (non-Indigenous owned Johnston's Cranberry Marsh), Iroquois Cranberry Growers engaged festival goers with cranberry bog tours, specialty products, and annual harvest information. In April 2017, Iroquois Cranberry Growers permanently closed its operation, a function of changing agricultural technologies and consistent over supply of cranberries within markets.[4]

The third case considers an apparent antithesis to nature tourism – *Casino Rama*. Located on reserve lands of the Chippewas of Rama First Nation, and just outside the city of Orillia in the heart of Ontario Lake Country, this entertainment and resort attraction is the result of a joint venture between the Chippewas of Rama, private, and government interests. Opening in 1996, Casino Rama is the only First Nations resort casino in Ontario and most successful Indigenous casino in Canada (Seaman, Prichard, & Potter, 2011). It generates annual revenues that reach over $500 million, employs approximately 700 Indigenous People, and has triggered infrastructure developments for the Chippewas of Rama, including a 300-room hotel, a spa, a health club, restaurants and bars, and upgrades to community schools, hospitals, water treatment, and housing facilities (Seaman et al., 2011). This success is due in large part to its proximity to Toronto and cottage country tourist markets, which Adese (2012) has persuasively argued are drawn in by the commoditization of indigeneity.

Methodology

Focusing on how Indigenous cultures are portrayed in the tourism promotional landscape of these case studies, we used critical discourse analysis to help expose and unsettle settler colonial power relations infiltrating tourism. While discourse analysis is practiced and conceptualized in diverse ways (Hannam & Knox, 2005), we understand it broadly as one of many possible approaches to textual analysis that draws on insights from ethnomethodology, poststructuralism, and sociolinguistics (Schwandt, 2015). Epistemological attachments direct the form that discourse analysis will take, and range from interpretivist and critical analyses of discourse as *dialogic linguistic exchanges of spoken word or text* to poststructural informed analyses of discourse as *power-laden ways of making sense of and engaging within the world* (Rose, 2001; Waitt, 2005). The epistemological positioning adopted in this paper takes direction from Foucault (1970) and poststructuralism in that we endeavored to analyze ways in which representations (language, texts, images) are produced by, and in turn productive of, dominant social dynamics and structures of power (e.g. settler colonialism). This approach is concerned with how certain discourses become solidified as being 'true,' while at once exploring how 'Othered' discourses are also imbued with power and work to subvert accepted regimes of truth (Waitt, 2005). Specific to colonial contexts, Bignall (2008) shows how engaging these analytics can enable colonizing peoples to critically examine 'historically imperial (and sometimes fascist) modes of being and social engagement … by drawing from alternative concepts of sociability already embedded within Western thought' (p. 129). Indeed, we apply such a critical analytical lens herein to render visible the relations of power that scaffold settler colonial narratives in tourism and potential avenues for different ways of knowing, and thus engaging with, the world around us.

Promotional texts incorporated into this study were compiled from websites associated with the case studies and a range of stakeholders with vested interests in the near north destination image. This process began with a Google search of keywords (e.g. 'Bala Cranberry Festival,' 'Ontario Lake Country,' or 'Algonquin Park + tourism'), which provided us with a general sense of promotional websites and online information. Next, we modified a purposive snowball sampling approach to identify, select, copy, and organize web-based promotions into Microsoft Excel spreadsheets. Beginning at a local scale, we extracted visual and textual information from official event or destination websites. Links from these official sites to other relevant websites were explored and mined for germane texts and imagery. The scope was then broadened to band, municipal, and regional levels, noting again any relevant links contained therein. Throughout this process, we were especially alert to including website information controlled by Indigenous communities, agencies, and operators. Tables 1–3 summarize the final list of website promotions integrated into our analysis.

Our analysis of compiled web content commenced with content analysis, which Rose (2001) indicates is a useful prelude to discourse analysis as it promotes familiarity with the manifest substance of texts being scrutinized. We started with multiple readings and viewings of promotional materials for the various ways Indigenous cultures were represented. Zachary then coded written website text using a line-by-line approach. Website images and videos were content analyzed by focusing on both manifest content, literally what is present, as well as latent content, consisting of the interpreted meanings and contexts (Mannay, 2016). Additionally, while working through the textual and visual materials, written memos were taken. Emphasis was placed on tracking researcher reflections, questions, observations of emerging patterns, and epiphanies relating to study objectives.

Meghan and Bryan then employed critical discourse analysis to read 'within' and 'beneath' the texts (Waitt, 2005). Given the significance of intertextuality in discourse analysis, we critically examined promotions for ways that particular discourses (e.g. those pertaining to nature, wilderness, land, history, and Indigeneity) are replicated and mutually reinforced within and across a range of texts (Waitt, 2005). Using as guideposts two principle questions – first, *how is settler occupation legitimized through promotions showcasing Indigenous culture?* and, second, *how might settler colonialism be 'undone' through promotions showcasing Indigenous culture?* – we searched for 'effects of truth'

Table 1. Analyzed promotional websites associated with Algonquin Park.

Organization/Operator	Type	Web Address*
Friends of Algonquin Park	Local Non-profit	http://www.algonquinpark.on.ca
Algonquin Park Residents Association	Local Non-profit	http://www.algonquinparkresidents.ca
Algonquin Outfitters	Local Private	http://algonquinoutfitters.com
The Portage Store	Local Private	https://portagestore.com
Arowhon Pines	Local Private	http://arowhonpines.ca
Bartlett Lodge	Local Private	http://www.bartlettlodge.com
Killarney Lodge	Local Private	http://www.killarneylodge.com
Explorer's Edge	Regional Government/Non-profit	http://explorersedge.ca/algonquin-canadas-iconic-park/
Ontario Highlights	Regional Public	http://ontarioshighlands.ca/discover/provincial-parks/algonquin-park/
Algonquins of Ontario	Regional Indigenous	http://www.tanakiwin.com/
Algonquin Canoe Company	Regional Private Indigenous	http://www.algonquincanoe.com
Ontario Parks	Provincial Public	https://www.ontarioparks.com/park/algonquin

*Our study examined several webpages accessed from the web addresses listed. All webpages were accessed between January 12 – February 02, 2017. Unless otherwise noted, websites are managed by non-Indigenous organizations and operators.

Table 2. Analyzed promotional websites associated with the Bala Cranberry Festival.

Organization/Operator	Type	Web Address*
Bala Cranberry Festival	Local Volunteer	http://www.balacranberryfestival.on.ca
Johnston's Cranberry Marsh	Local Private	http://www.cranberry.ca/Farm/home.html
Muskoka Lakes Winery	Local Private	http://cranberry.ca/Winery/festival.html
Iroquois Cranberry Growers	Local Indigenous	http://iroquoiscranberries.com
Bala Muskoka Ontario	Local Public	http://www.bala.muskoka-ontario.com/index.html
Muskoka Lakes Chamber of Commerce	Regional Non-profit	http://www.muskokalakeschamber.ca
District of Muskoka	Regional Government	http://www.muskoka.on.ca/en/index.aspx
Muskoka Tourism	Regional Non-profit	http://www.discovermuskoka.ca
Explorer's Edge	Regional Government/Non-profit	http://explorersedge.ca

*Our study examined several webpages accessed from the web addresses listed. All webpages were accessed between May 26 – July 15, 2016. Unless otherwise noted, websites are managed by non-Indigenous organizations and operators.

that shape discourse, as well as silences appearing as gaps in the promotional landscape (Rose, 2001; Waitt, 2005). Analysis proceeded through additional readings/viewings of the website materials, coding illustrative 'moments' in the text, reviewing codes and themes from the content analysis, and writing. Throughout this process, we consistently and reflexively questioned what these texts aim to *do*, the power relations within which they were produced, as well as our own understandings of discourse, settler colonialism, and tourism in the near north. In this sense, our positionality as analysts and authors (i.e. as Settler Canadians of European descent) is embedded in findings that follow. Our intentions are, therefore, not to establish any final 'truth' but rather to reveal how we see discourse being constituted by and (re)constructive of power relations in society.

Findings

The outcomes of our analysis are presented in two parts. First, we delve into how settler colonial occupation is legitimized within and through the promotional terrains of tourism in Ontario's

Table 3. Analyzed promotional websites associated with Casino Rama.

Organization/Operator	Type	Web Address*
Casino Rama Resort	Local Private	https://www.casinorama.com
Chippewas of Rama First Nation	Local Indigenous	http://www.ramafirstnation.ca
The Gathering Place	Local Private	http://www.thegatheringplacerama.com/
City of Orillia	Local Public	http://www.orillia.ca/en/
Township of Oro-Medonte	Regional Government	http://www.oro-medonte.ca
Ontario's Lake Country	Regional Non-profit	http://www.ontarioslakecountry.com
Tourism Simcoe Country	Regional Industry/Non-profit	http://experience.simcoe.ca
Regional Tourism Organization 7	Regional Government/Non-profit	http://brucegreysimcoe.com/home

*Our study examined several webpages accessed from the web addresses listed. All webpages were accessed between May 27 – July 26, 2016. Unless otherwise noted, websites are managed by non-Indigenous organizations and operators.

near north. Second, we illuminate how Indigenous cultural relations are being asserted through tourism promotions, which is suggestive of how settler colonial narratives can be undone.

Legitimizing settler occupation

Even the most cursory glance at the promotional landscapes of Algonquin Park, the Bala Cranberry Festival, and Casino Rama yields a sense of the prevailing discourse that circulates within the production of tourism in Ontario's near north. It is one tethered to ideologies and imaginaries of 'nature' as an untouched and stunning backdrop for pleasure-seeking pursuits and fulfilling leisure desires. Phrases such as 'cherished and pristine natural environment,[5]' 'unspoiled lakes and rivers[6],' or a destination 'waiting for you to explore[7]' appear frequently, and are coupled with images of wild granite shorelines, lush forests, and picturesque waterways. When populated with people, images default to showcasing various forms of outdoor recreation – boating, canoeing, wakeboarding, relaxing on a cottage dock – suggesting that the near north is meant to be experienced and consumed as nature-based getaway. Website promotions of Algonquin Park are especially indicative here as audiences are consistently informed that the park contains 'some of the world's most beautiful scenery[8],' 'picture-post card vistas[9],' and 'legendary wildlife[10]' such as moose, loons, wolves, and bear. Visitors are therefore persuaded to identify the Park as a 'paradise for nature lovers[11]' and a place that 'offers an unparalleled wilderness experience[12]'. Ontario's near north is thus enfolded within broader productions and representations of 'wilderness' (Braun, 2002; Grimwood et al., 2015) and the 'Great Outdoors'. Even Casino Rama plays on this discourse of nature by inviting guests to play and be entertained in 'The Great Indoors[13]'. Innocent though they may seem, such promotions are imbued with power to shape destination places by dictating who belongs and what behaviors and values are appropriate, and conversely, who and what are deemed out-of-place (Morgan & Pritchard, 1998). In Canada (though not exclusively), it has been shown that nature spaces are discursively scripted by and for the colonizing gaze and consumption of white, masculine, and heterosexual audiences (Erickson, 2018; Pritchard & Morgan, 2000; Yudina, Grimwood, Berbary, & Mair, 2018). Here, 'nature' represents a primordial physical realm that is at once distinct from, but threatened by, modern civilization, save of course for well-behaving nature enthusiasts who are often nostalgic for a more organic, premodern existence (Braun, 2002). 'Traditional' Indigenous ways of being – those represented as autochthonous and belonging to time that once was – are deemed appropriate in such spaces (Erickson, 2018; Grimwood et al., 2015). The contemporary and *living* presence of Indigenous Peoples becomes obscured, or dismissed entirely as being 'unauthentic' or 'tainted' by modernity (Braun, 2002).

In addition to constructing the near north as a pristine nature-scape for visitors and second-home owners from metropolitan areas to the south, the tourism promotional materials reinforce and legitimize a number of 'settler common sense' strategies (to borrow this phrasing from Rifkin, 2013). The first of these strategies evokes a simpler time, a return to traditional family values and living in harmony with the land. These settler stories speak of long attachments to the region and generations of presence and stewardship of the landscape. Textual and visual representations in the promotional discourse establish the benefits that settlers have brought and continue to bring to the region, and how visitors to the near north can partake in and witness this simple lifestyle that has remained virtually unchanged over time. Johnson's Cranberry Marsh, for instance, shares black and white photographs of family ancestors and an old 1940s Dodge pickup on their website. The website promotes the agricultural work of the family, but also evokes a historicized connection to place; a modest, authentic lifestyle that tourists can encounter and experience. This is in contrast to the Wahta Mohawk's website, which presented their operations in more agricultural, economic, and community development terms. In Bala, a number of websites speak to 'the cranberry tradition' and the history of family farming; however, the town website only acknowledges the settler history of the area, and none speak to any contemporaneous relationships with the cranberry farm run by the Wahta Mohawk First Nation.

The Algonquin Park Residents Association (APRA), 'a small group of leasehold families, who have occupied small corners of the Park since the earliest days of the twentieth century[14],' works to firmly situate themselves as rightful inhabitants using a similar discursive strategy to historically validate settler presence. This is evident in how they speak of coming to reside in the Park, firmly establishing the date of their arrival, as well as making reference to 'the fifth generation of leaseholders' whose 'forebears were invited in the early years of the twentieth century, by the Ontario Government, to establish cottages in the Park[15].' By representing their 'right' to be present in the Park, and specifically referencing the government's involvement in this process, the APRA elides any dispute of their occupation of the space and legitimizes their ways of being on the land. The APRA further justifies their inhabitation through the minimization of their environmental footprint in the Park. Phrases such as 'a small number of cabins sheltered amid the darkness of the local forest[16],' and 'in these cabins live quiet, unobtrusive groups of leaseholders and their extended families,[17]' suggest a virtually invisible impact in the Park. This speaks to previous research which explored how the ethic of minimal ecological footprints in spaces deemed as 'pristine' served to privilege 'responsible' activities on the landscape, while at once delegitimizing contemporary Indigenous settlements or livelihood activities (Grimwood et al., 2015).

The promotional materials further legitimize settler narratives through the ways in which they tend to code Indigenous knowledge as 'myth' or 'legend.' We found only one reference to Indigenous antecedence in the cranberry production history of Bala:

> The natives who discovered cranberries used them for food, medicine and dye. Cranberries were so important, myths were created. According to one tribe, cranberries were a gift from the Great Spirit brought to earth in the beak of a dove.[18]

The use of the word 'myth' to describe the story of the cranberry undermines Indigenous knowledge systems. Employing this language functions as a dismissal of Indigenous understanding and speaks to the systemic erasure of Indigenous presence, while at once serving to legitimize the modern, yet historically-established and family-based, approach to farming adopted by settlers. Establishing Indigenous understandings of farming and their approach to land stewardship as Other, counter to 'normalized' and 'modern' farming practices, infantilizes Indigenous knowledges and their relationships with the land. Hollinshead (1992, 1998) issued similar critiques of the ways in which mainstream culture seeks to mythologize and homogenize Indigenous beliefs and practices in the United States. Salazar (2013) has likewise argued that Indigenous Peoples and their traditions are stereotypically represented as a series of 'mythologized and eroticized imaginaries' (p. 34).

One of these romanticized imaginaries is the characterization of Indigenous Peoples as nomadic, unfettered by and inferior to settler concepts such as property ownership and place attachment. Discursively constructing nomadic land relations as 'less than' de legitimizes Indigenous Peoples' genuine claims to being original and now displaced occupants of the land. The only mention of Indigenous antecedence on the Friends of Algonquin website is at the very beginning of their 'Cultural History' page, where it is written:

> For most of Algonquin's history, human settlement was not a very important element. Scattered family groups of aboriginal Peoples came to fish, hunt and pick berries, but their numbers were never large. It was not until the 1800s that big changes came to the rugged Algonquin highlands.[19]

The discourse here implies that the land truly became occupied with the advent of the logging industry and the development of the Park. This language works further to affirm capitalistic notions of property ownership and inhabitation, wherein being an owner/inhabitant means residing in one place, a value-based assumption that does not necessarily accord with Indigenous understandings of land stewardship.

Another strategy that legitimizes settler colonialism identified in the promotional landscape is the way in which Indigenous displacement is characterized as a function of tribal conflict. Located in Algonquin Park, Bartlett Lodge acknowledges Algonquin Peoples as the original inhabitants of the Park; however, they characterize their displacement as a result of conflict amongst Indigenous groups:

> The Algonquins suffered near annihilation by the Iroquois. Victims of European politics, the Algonquins were driven from their vast lands by the Iroquois Confederacy. The powerful Iroquois, aided first by the Dutch and later by the English, defeated both the French and Algonquins.[20]

This statement grossly diminishes the European role is destabilizing Indigenous communities and fomenting conflict amongst the Indigenous inhabitants of the region. More subtly, it also works to align certain European parties as victims. Establishing a distance from historical complicities, while at once affirming a desire for the well-being of Indigenous Peoples, are central strategies in validating and rendering invisible the settler colonial project (Cooke, 2017).

A related settler strategy is to trivialize the seriousness and ongoing effects of Indigenous displacement. Indeed, a great irony of settler occupation occurs when Indigenous erasure reaches such an extent that it makes sense for settlers to desire and lament Indigenous cultures and traditions. We read this in the content on Bartlett Lodge's website which speaks of 'things hav[ing] a way of coming full circle.[21]' One might assume that by 'full circle' they mean that Algonquin presence on the land has been restored; however, the website goes on to say:

> The Algonquin People, the first inhabitants of the area used tents for shelter. Today, guests at Bartlett Lodge in Algonquin Park can also experience the wonder of sleeping under canvas, in decidedly the most luxurious style of camping; platform tenting.[22]

By insinuating that things are back to the way they once were on the land, this passage contributes to a discourse of devaluing the Algonquin claims to land and the generations of trauma that they have endured as a result of the colonial presence. The historicization of Indigenous Peoples, accompanied as it is by the tropes of the long presences and small footprints of the settlers serves to effectively erase contemporary Indigenous rights to place while firmly embedding the legitimacy of the settlers to be there.

Asserting Indigenous cultural relations

This section turns attention to how settler colonialism might be undone through promotions showcasing Indigenous culture. As we flagged earlier, several recent studies have illustrated how Indigenous Peoples are harnessing the socio-cultural, economic, and political forces of tourism in order to disrupt and challenge prevailing representations of Indigeneity in tourism generally (Butler & Hinch, 2007; Holmes et al., 2016) and tourism promotions specifically (Amoamo & Thompson, 2010; Seiver & Matthews, 2016). In this section, we reveal ways in which Indigenous stakeholders in Ontario's near north are engaging in similar politics.

Against the strong trend in the tourism promotional material towards historicizing Indigenous Peoples, a counter-discourse revealed in our analysis had to do with showcasing contemporary Indigenous livelihoods. For instance, the website for the Iroquois Cranberry Growers portrays cranberry farming as a means for maintaining a self-sufficient local economy. As a community-owned business, the Iroquois Cranberry Growers presents itself as a 'project designed to provide employment for community members, and to support an economic base for the reserve government.[23]' It is presented less as a destination for tourists to consume nature or Indigeneity than it is as a place where modern Indigenous agriculture fosters community identity and economic wellbeing. The scarce mention of the Bala cranberry festival on the Iroquois Cranberry Growers webpages seems to mirror this indifference to tourism. *trawlwulwuy* scholar, Emma Lee (in press) of *tebrakunna* country (Tasmania) has recently observed how tourism is approached similarly by Basque women operating fishing collectives in northeastern Spain. When cultural safety and livelihoods are the core of community concerns, Lee argues that tourism is rarely the panacea it is made out to be and should be viewed as, at best, supplemental. If as a field we are serious about disrupting (settler) colonialism in and through tourism, we must be open to the very real possibility that sometimes – for Indigenous Peoples and other historically marginalized groups subjected to our gaze – tourism does not matter.

On the website of the Chippewas of Rama First Nation, we observed no images of Indigenous Peoples in so-called 'traditional dress.' Instead, website imagery portrayed soldiers, policemen, athletes, children playing sports, and people gathering in community spaces. These images are indicative of the multiple and contemporary roles, experiences, and livelihoods of those living in the community. Approached in relation to tourism, they work to counter the prevailing tendency to historicize – and, in so doing, remove from the discourses of modernity – Indigenous Peoples (Braun, 2002). Moreover, the depiction of modern Chippewas of Rama citizens on these webpages jars the conflation of authentic Indigeneity with traditional (i.e. 'primitive'), nature-based lifestyles often depicted in tourism and which disciplines tourist desires (Braun, 2002; Grimwood et al., 2015; Jasen, 1995). Indeed, this presentation of a contemporary Indigenous Peoples stands as a direct challenge to the essentialism of Indigenous identity and culture that settler colonialism pursues (Lowman & Barker, 2015).

In conjunction with representations of the Chippewas of Rama First Nation as active members in a recognizable contemporary society, promotional websites also highlight the importance of acknowledging the past, saying, for instance, 'our First Nation community is proud of our history, Ojibway language and culture.[24]' This text goes on to assert a community identity of a 'caring, sharing people ... known for [their] hospitality[25]' and whose territory has become synonymous with 'the gathering place[26]', which is explained to be rooted in their historical relationships with other First Nation communities and early Settlers. Clearly, and in direct contrast text cited earlier from the Friends of Algonquin Park website, inter-tribal violence and conflict was not the sole mode of relationship between Indigenous Peoples. Indeed, as Chippewas of Rama establish, Indigenous Peoples had often peaceable relations with one another prior to European arrival, thus disrupting the deeply rooted settler notion of Indigenous Peoples as warring 'savages' and in need of the 'civilizing' influence of Europeans (Lowman & Barker, 2015).

In the discussion of their history, the Chippewas of Rama First Nation also reflect on themes of loss and resilience. For example, one passage reads:

> Around 1830, our community was moved to the Coldwater Narrows area by the Crown, part of an 'experiment' which shaped 'Indian Reserves.' We continued on as industrious people, building a road for commerce which is known today as Highway 12, establishing farms, mills, and markets for selling produce, fish and game to settlers and travellers.[27]

The use of quotation marks around the words 'experiment' and 'Indian Reserves' insinuates a challenge to the legitimacy of these words. This discourse of loss and resilience is reinforced when they speak of being 'forced to move again after [their] land was taken,' and how because their new 'land was difficult to farm' and they had 'los[t] [their] inherent right to fish and hunt with the disputed Williams Treaty in 1923, [they] pursued other entrepreneurial opportunities in the tourism market,[28]' including the establishment of Casino Rama. Turning to tourism in this case served as a continuation of Chippewas of Rama's innovative responses and adaptations to the unrelenting burdens of settler colonialism. Perhaps the commoditization of Indigeneity (Adese, 2012) that accompanies Casino Rama is accepted by the community as a necessary transgression to ensure the continued success of the operation.

The notions of loss and resilience are also present in Algonquin Peoples' efforts to counter their erasure from contemporary livelihoods in Algonquin Park. As the Algonquins of Ontario explain, after inter-generational trauma at the hands of colonizers, '[We] are on a journey of survival, rebuilding, and self-sufficiency – a journey of reconciliation.[29]' They further illuminate their perseverance in relation to political recognition and land claims: 'Algonquin petitions to the Crown seeking recognition and protection for Algonquin land and other rights date back to 1772.[30]' Their continued work towards 'reconciliation and renewed relationships[31]' was embodied in a totem pole hand-carved by an Algonquin elder residing east of the park. On behalf of the Algonquins of Ontario, the totem pole was gifted to and given a home in Algonquin Park, '[i]n the spirit of peace and reconciliation with all nations, as well as a way to show respect to all cultures.[32]' Related moves are being made by Algonquin Canoe Company, which strives to return Indigenous practices and traditions to contemporary ways of being in the Park. As their website explains, '[s]ince time before memory, the traditional waterways have been the heart and bloodline of the Algonquin People.[33]' Trade roots have been replaced with paddle routes, which offers a way to continue to preserve and celebrate traditional knowledge and cultural histories. Counter to the silencing of Indigenous presences in other promotional materials, the Algonquin Canoe Company assures readers that the 'Algonquin People used the Park area for thousands of years before its establishment.[34]' These actions serve to embed Indigenous controlled meanings, representations, and practices within the park landscape (see also Holmes et al., 2016).

Conclusion

In this paper, we have taken up the recurrent task of unmasking oppressive discourses that silence alternative ways of *doing* or *thinking about* tourism (Hollinshead, 1992). In particular, our critical discourse analysis of web-based tourism promotions associated with three sites in Ontario's near north exposed how tourism can circulate and authorize settler colonial power and occupation of Indigenous lands. We also revealed ways that Indigenous Peoples and communities involved in this promotional landscape are resisting the erasure and essentialism of culture that settler colonialism pursues.

Tourism, as Cooke (2016) argued, must be troubled as a space where settler colonial power produces particular 'truths' about Indigenous identities, cultures, and lands. At stake here are the politics of who controls the production of touristic and toured identities and places, how these become desired and consumed, and the consequences this all has in the lives of Indigenous Peoples and Settlers (Amoamo & Thompson, 2010). Ultimately, our analyses show how promotions associated with Settler operators, agencies, and institutions make legible and legitimate Settler heritage in the near north on the one hand, and serve to obscure, historicize, and essentialize Indigenous cultures on the other. While these are not novel insights within the wider tourism literature, they are interpretations contextualized within a particular destination that is not only helping to naturalize the 'logic of elimination' that underpins settler colonialism (Wolfe, 2006) but also seemingly reliant upon it. They therefore demonstrate how tourism is contributing to the settler colonial present; that is, to the structures and stories that make settler colonial invasion an ongoing form of occupation (Veracini, 2015). By exposing these settler colonial power relations in the near north's promotional landscape, our paper helps to disturb their invisibility and show, once again, how the pleasures of tourism are often coupled with grave social consequence (Caton, 2012).

Colonial discourses, as Braun (2002) argued, can be perforated. On a theoretical level, they must be, lest we accept them as totalizing and therefore inescapable. As our analyses revealed, Indigenous stakeholders in the near north appear to be engaging the promotional landscape in ways that destabilize norms or 'truths' of settler colonialism. The plurality within Indigenous cultures, livelihoods, and relations to land illustrated in our analyses undercut settler structures and stories that erase, anachronize, and essentialize Indigeneity, or use it entirely to turn a profit. Similar to what others have observed in Maori and Indigenous Australian contexts (Amoamo & Thompson, 2010; Seiver & Matthews, 2016), Indigenous Peoples and communities that control how they are represented in the near north do so in ways that speak to troubled histories and hopeful futures, land rights and stewardship over time, the complex and dynamic nature of identity, and the resiliency and relationships that enable Indigenous cultures to be sustained. The stories resist settler colonial ones flowing through much tourism and suggest alternative – possibly anti-colonial or decolonizing – ways of relating to tourism places in the near north.

This raises an important question about the role that Settlers can play in disrupting tourism's investments in the settler colonial present and future. For too long, Indigenous Peoples have shouldered disproportionate levels of responsibility for resisting the many faces of colonialism. This is a trend we observe echoing in much Indigenous tourism literature, and certainly apparent in the promotional landscapes featured in our research. Learning from Carlson (2017) and Lowman and Barker (2015), however, we would argue that it is our responsibility as Settlers to push back against the structures, systems, discourses, and narratives of settler colonialism, and the disavowal that helps define our identity. If, as Regan (2010) aptly notes, settler colonialism is about *taking space* from Indigenous peoples, unsettling settler colonialism must be about *making space*. We can best contribute to this not by continuously working to 'improve' or create 'better' representations of Indigenous culture – Indigenous Peoples and communities ought to control these – but by asking ourselves, other Settlers, and our institutions uncomfortable questions about our shared complicity in settler colonialism, making known the recurrent violence of settler colonialism and its conceptual baggage (e.g. *terra nullius*), and learning about the specificity and vitality of Indigenous cultures so that we are better prepared to relate respectfully and responsibly.

Canada's Truth and Reconciliation Commission has shown that respectful intercultural dialogue and learning is needed to inform and reframe Canadian consciousness (TRC, 2015). Tourism provides a lens through which to study these intersections of culture and the roles of Indigenous and Settler actors in the shaping of space, place, and identity. Indeed, tourism is a potent social force with the potential to both bridge or bolster cultural divides between these groups (Cooke, 2017; Higgins-Desbiolles, 2003). Our critically informed research has advanced understanding of how contemporary tourism can facilitate or impede social aspirations of reconciliation, challenging tourism discourse to shift away from settler concepts and discourse. It also contributes to the growing body of literature critiquing settler colonial narratives in the context of tourism, and adds to the discussion regarding the potentialities to be found at the nexus between Indigenous cultural heritage and tourism in settler colonial contexts. There is, of course, much more to be done. Given our limited focus on promotional landscapes, future research must consistently interrogate how settler colonial oppressions are embedded within policy, development, and experiential contexts of tourism, and allocate consistent resources towards the action of undoing them.

Notes

1. Indigenous Peoples in settler colonial Canada are incredibly diverse and include First Nations, Inuit, and Métis Peoples. First Nation is 'a term that came into common usage in the 1970s to replace the word 'Indian', which some people found offensive' (CIHR et al., 2014, p. 136).
2. An estimated 150,000 children were taken from their families and placed in residential schools, many of them being subjected to physical, emotional, verbal, and sexual abuse (TRC, 2015). Much has been written on this chapter of Canadian nation-making and we have no space to review it here. See Regan (2010) for an influential and related source.
3. Personal communication with Wahta Mohawk Economic Development Officer, May 2017.
4. Ibid.
5. The District Municipality of Muskoka. (2016).
6. Algonquin Outfitters. (2017).
7. Muskoka Lakes Chamber of Commerce. (2016).
8. The Portage Store. (2017).
9. Bartlett Lodge. (n.d.).
10. Arowhon Pines. (n.d.).
11. Ibid.
12. Ontario Parks. (2017).
13. Casino Rama Resort. (2017).
14. Algonquin Park Residents Association. (n.d.a).
15. Algonquin Park Residents Association. (n.d.b)
16. Ibid.
17. Ibid.

18. Johnston's Cranberry Marsh. (2016).
19. Friends of Algonquin Park (2017).
20. Bartlett Lodge. (n.d.).
21. Ibid.
22. Ibid.
23. Iroquois Cranberry Growers. (2016).
24. Chippewas of Rama First Nation. (n.d.b).
25. Chippewas of Rama First Nation. (n.d.a).
26. Ontario's Lake Country. (2016).
27. Chippewas of Rama First Nation. (n.d.b).
28. Chippewas of Rama First Nation. (n.d.a).
29. Algonquins of Ontario. (2013d).
30. Algonquins of Ontario. (2013b).
31. Algonquins of Ontario. (2013a).
32. Algonquins of Ontario. (2013c).
33. Algonquin Canoe Company. (2015b).
34. Algonquin Canoe Company. (2015a).

Disclosure statement

No potential conflict of interest was reported by the authors.

Funding

This work was supported by Ontario Ministry of Research, Innovation and Science [Early Researcher Award, project number ER15-11-182].

ORCID

Bryan S. R. Grimwood http://orcid.org/0000-0003-1555-7541

References

Adese, J. (2012). *Aboriginal™: Constructing the aboriginal and imagineering the Canadian national brand (Unpublished doctoral dissertation)*, McMaster University.

Aitchison, C., MacLeod, N., & Shaw, S. (2000). *Leisure and tourism landscapes: Social and cultural geographies*. New York: Routledge.

Algonquin Canoe Company. (2015a). *Tour: Algonquin Park*. Retrieved from http://www.algonquincanoe.com/algonquin-park/

Algonquin Canoe Company. (2015b). Welcome to the Algonquin canoe company. Retrieved from http://www.algonquincanoe.com

Algonquin Outfitters. (2017). *Canoe trips*. Retrieved from http://www.algonquinoutfitters.com/canoe-trips/

Algonquin Park Residents Association. (n.d.a). *History*. Retrieved from http://www.algonquinparkresidents.ca/history.html

Algonquin Park Residents Association. (n.d.b). *Who are Algonquin leaseholders?* Retrieved from http://www.algonquinparkresidents.ca/leaseholders.html

Algonquins of Ontario. (2013a). Algonquins of Ontario. Retrieved from http://www.tanakiwin.com

Algonquins of Ontario. (2013b). *Overview of treaty negotiations*. Retrieved from http://www.tanakiwin.com/our-treaty-negotiations/overview-of-treaty-negotiations/

Algonquins of Ontario. (2013c). Totem pole in Algonquin Park. Retrieved from http://www.tanakiwin.com/current-initiatives/current-initiativestotem-pole-in-algonquin-park/

Algonquins of Ontario. (2013d). *Who are the Algonquins of Ontario?* Retrieved from http://www.tanakiwin.com/algonquins-of-ontario/who-are-the-algonquins-of-ontario/

Amoamo, M., & Thompson, A. (2010). (Re)Imaging Māori tourism: Representation and cultural hybridity in postcolonial New Zealand. *Tourist Studies*, *10*(1), 35–55.

Arowhon Pines. (n.d.). *Animals*. Retrieved from https://www.arowhonpines.ca/play/animals/

Bartlett Lodge. (n.d.). *Lakeside, Algonquin Park, Ontario resort*. Retrieved from http://www.bartlettlodge.com

Bennett, N., Lemelin, R. H., Johnston, M., & Koster, R. L. (2010). Using the social economy in tourism: A study of National Park creation and community development in the Northwest Territories, Canada. *Journal of Rural and Community Development*, *5*(1), 200–220.

Bignall, S. (2008). Postcolonial agency and poststructuralist thought: Deleuze and Foucault on desire and power. *Angelaki-Journal of the Theoretical Humanities*, *13*(1), 127–147.

Braun, B. (2002). *The intemperate rainforest: Nature, culture, and power on Canada's West Coast*. Minneapolis: University of Minnesota Press.

Butler, R., & Hinch, T. (2007). *Tourism and Indigenous Peoples: Issues and implications*. New York: Elsevier.

Carlson, E. (2017). Anti-colonial methodologies and practices for settler colonial studies. *Settler Colonial Studies*, *7*(4), 496–517.

Casino Rama Resort. (2017). *Home*. Retrieved from https://www.casinorama.com/

Caton, K. (2012). Taking the moral turn in tourism studies. *Annals of Tourism Research*, *39*(4), 1906–1928.

Chippewas of Rama First Nation. (n.d.a). *About us*. Retrieved from http://www.ramafirstnation.ca/about.asp

Chippewas of Rama First Nation. (n.d.b). *Chippewas of Rama First Nation owned businesses*. http://www.ramafirstnation.ca/current.asp

Canadian Institute of Health Research, Natural Sciences and Engineering Research Council of Canada, & Social Sciences and Humanities Research Council of Canada (CIHR, NSERC, & SSHRC). (2014, December). *Tri-Council policy statement: Ethical conduct for research involving humans*. Ottawa.

Cohen, E. (1993). The study of touristic image of native people: Mitigating the stereotype of a stereotype. In D. G. Pearce, & R. Butler (Eds.), *Tourism research critiques and challenges* (pp. 36–69). London: Routledge.

Cooke, L. (2016). 'North' in contemporary Canadian national-cultural imaginaries: A haunted phantasm. *Settler Colonial Studies*, *6*(3), 235–251.

Cooke, L. (2017). Carving "turns" and unsettling the ground under our feet (and skis): A reading of Sun Peaks resort as a settler colonial moral terrain. *Tourist Studies*, *17*(1), 36–53.

Coulthard, G. (2014). *Red skin, white masks*. Vancouver: UBC Press.

The District Municipality of Muskoka. (2016). *Live and play*. Retrieved from http://www.muskoka.on.ca/en/live-and-play/Live-and-Play.aspx

Erickson, B. (2018). Embedded dangers and anachronistic others. In B. Grimwood, K. Caton, & L. Cooke (Eds.), *New moral natures in tourism*. New York: Routledge.

Foucault, M. (1970). *The order of things: An archaeology of the human sciences*. New York: Vintage Books.

Friends of Algonquin Park. (2017). *Cultural history*. Retrieved from http://www.algonquinpark.on.ca/visit/history/cultural-history.php

Gehl, L. (2013). The basic distinctions: Anishinaabe and Haudenosaunee. *Black face blogging*. Retrieved from https://www.lynngehl.com/black-face-blogging/the-basic-distinctions-anishinaabe-and-haudenosaunee

Grimwood, B. S. R., Yudina, O., Muldoon, M., & Qiu, J. (2015). Responsibility in tourism: A discursive analysis. *Annals of Tourism Research*, *50*, 22–38.

Hall, M. C., & Tucker, H. (2004). *Tourism and postcolonialism: Contested discourses, identities, and representations*. New York: Routledge.

Hannam, K., & Knox, D. (2005). Discourse analysis in tourism research: A critical perspective. *Tourism Recreation Research*, *30*(2), 23–30.

Harrison, J. (2013). *A timeless place: The Ontario cottage*. Vancouver: UBC Press.

Higgins-Desbiolles, F. (2003). Reconciliation tourism: Tourism healing divided societies. *Tourism Recreation Research*, *28*(3), 35–44.

Hollinshead, K. (1992). 'White' gaze, 'red' people – shadow visions: The disidentification of 'Indians' in cultural tourism. *Leisure Studies*, *11*(1), 43–64.

Hollinshead, K. (1998). Tourism, hybridity, and ambiguity: The relevance of Bhabha's "third space" cultures. *Journal of Leisure Research*, *30*(1), 121–156.

Holmes, A., Grimwood, B. S. R., King, L., & the Lutsel K'e Dene First Nation. (2016). Creating an indigenized visitor code of ethics: The development of Denesoline self-determination for sustainable tourism. *Journal of Sustainable Tourism*, *24*(8–9), 1177–1193.

Iroquois Cranberry Growers. (2016). Harvest. Retrieved from http://iroquoiscranberries.com/harvest/

Jasen, P. (1995). *Wild things: Nature, culture, and tourism in Ontario, 1790–1914*. Toronto: University of Toronto Press.

Johnston's Cranberry Marsh. (2016). *Cranberry facts*. Retrieved from http://cranberry.ca/johnstons-cranberry-marsh/cranberry-facts/

Lee, E. (2017). Performing colonisation: The manufacture of Black female bodies in tourism research. *Annals of Tourism Research, 66*, 95–104.

Lee, E. (in press). Black female cultural safety in tebrakunna country: What is wellness for us? In B. S. R. Grimwood, H. Mair, K. Caton, & M. Muldoon (Eds.), *Tourism and wellness: Travel for the good of all?* Lanham, MD: Lexington.

Lowman, E. B., & Barker, A. (2015). *Settler: Identity and colonialism in 21st century Canada*. Halifax: Fernwood Publishing.

Mannay, D. (2016). *Visual, narrative and creative research methods: Application, reflection and ethics*. New York, NY: Routledge.

Morgan, N., & Pritchard, A. (1998). *Tourism promotion and power: Creating images, creating identities*. Chichester: John Wiley & Sons Ltd.

Muskoka Lakes Chamber of Commerce. (2016). *Welcome*. Retrieved from http://www.muskokalakeschamber.ca/welcome

Notzke, C. (1999). Indigenous tourism development in the Arctic. *Annals of Tourism Research, 26*(1), 55–76.

Ontario, Province of. (2018). Retrieved from https://www.ontario.ca/page/algonquin-land-claim#section-3

Ontario's Lake Country. (2016). *Chippewas of Rama First Nation*. Retrieved from http://www.ontarioslakecountry.com/listing/chippewas-rama-first-nation

Ontario Parks. (2017). Algonquin: Backcountry. Retrieved from https://www.ontarioparks.com/park/algonquin/backcountry

The Portage Store. (2017). *Wilderness canoe camping*. Retrieved from https://www.portagestore.com/canoeing/wilderness-canoe-camping/

Pritchard, A., & Morgan, N. (2000). Privileging the male gaze: Gendered tourism landscapes. *Annals of Tourism Research, 27*(4), 884–905.

Ray, A. (2016). *An illustrated history of Canada's native people* (4th ed). Montreal & Kingston: McGill-Queen's University Press.

Regan, P. (2010). *Unsettling the settler within*. Vancouver: UBC Press.

Rifkin, M. (2013). Settler common sense. *Settler Colonial Studies, 3*(3–4), 322–340.

Rose, G. (2001). *Visual methodologies: An introduction to the interpretation of visual materials*. London: SAGE Publications.

Ryan, C., & Aicken, M. (2005). *Indigenous tourism: The commodification and management of culture*. Amsterdam: Elsevier.

Salazar, N. B. (2013). The (im)mobility of tourism imaginaries. In M. Smith, & G. Richards (Eds.), *The Routledge handbook of cultural tourism* (pp. 34–39). New York, NY: Routledge.

Schwandt, T. A. (2015). *The Sage dictionary of qualitative inquiry* (4th ed.). Thousand Oaks, CA: Sage.

Seaman, P., Prichard, B., & Potter, D. (2011). Betting on reconciliation: Law, self-governance, and First Nations economic development in Canada. *Gaming Law Review and Economics, 15*(4), 207–219.

Seiver, B., & Matthews, A. (2016). Beyond whiteness: A comparative analysis of representations of aboriginality in tourism destination images in New South Wales. *Australia. Journal of Sustainable Tourism, 24*(8–9), 1298–1314.

Truth and Reconciliation Commission of Canada(TRC). (2015). *Honouring the truth, reconciling for the future: Summary of the final report of the Truth and Reconciliation Commission of Canada*. Library and Archives Canada.

Urry, J. (1992). *The tourist gaze*. Thousand Oaks, CA: Sage.

Veracini, L. (2011). Introducing settler colonial studies. *Settler Colonial Studies, 1*(1), 1–12.

Veracini, L. (2015). *The settler colonial present*. New York: Palgrave Macmillian.

Viken, A., & Müller, D.2017). *Tourism and indigeneity in the Arctic*. Blue Ridge Summit, PA: Channel View Publications.

Waitt, G. (2005). Doing discourse analysis. In I. Hays (Ed.), *Qualitative research methods in human geography* (pp. 163–191). London, UK: Oxford University Press.

Walker, K., & Moscardo, G. (2016). Moving beyond sense of place to care of place: The role of Indigenous values and interpretation in promoting transformative change in tourists' place images and personal values. *Journal of Sustainable Tourism, 24*(8–9), 1243–1261.

Whitford, M., & Ruhanen, L. (2016). Indigenous tourism research, past and present: Where to from here? *Journal of Sustainable Tourism, 24*(8–9), 1080–1099.

Wolfe, P. (2006). Settler colonialism and the elimination of the native. *Journal of Genocide Research, 8*(4), 387–409.

Yang, L. (2011). Ethnic tourism and cultural representation. *Annals of Tourism Research, 38*(2), 561–585.

Yudina, O., Grimwood, B. S. R., Berbary, L., & Mair, H. (2018). The gendered natures of polar bear tourism. *Tourism Culture and Communication, 18*(1), 51–66.

OPEN ACCESS

Authenticity as a compromise: a critical discourse analysis of Sámi tourism websites

Cecilia de Bernardi

ABSTRACT

Authenticity is considered a very important means to attract tourists. Different enterprises and destinations compete for visitors and authenticity is an important factor for visitors' travel decisions. Within the context of indigenous and cultural tourism, several studies have criticized the way in which indigenous and local populations are portrayed in order to attract tourists, since they often involve colonial images and descriptions that are no longer corresponding to the contemporary situation. This study presents an alternative perspective on the discussion on tourism marketing in relation to authenticity. Through critical discourse analysis and retroduction, this study has analysed the websites of Sámi tourism companies from Sweden. The discussion put forward in this study is not a generalizing picture of how the Sámi peoples choose to market themselves, but it problematizes how these particular enterprises relate to issues of authenticity in the representation of cultural heritage. This paper also presents a new approach to authenticity seen as a compromise. The results of this study present one side of a multifaceted discussion on the struggle between different discourses on representations and authenticity that are often the main channel to reach and attract potential visitors.

Introduction

The Sámi are one of the indigenous populations of Europe. They live in a territory that extends over Norway, Finland, Sweden, and Russia. Although the exact Sámi population is not known, the International Work Group for Indigenous Affairs (http://www.IWGIA.org) estimates it to be between 50,000 and 100,000. Their territory is collectively called *Sápmi*. The Sámi also speak different languages in various regions, which are not always mutually understandable. There are other substantial differences between Sámi located in different areas. As Müller and Viken (2017, p. 3) argue: 'Tourism is basically founded on differences.' The Sámi in Scandinavia have been involved in tourism for many decades, with both negative and positive effects: 'Indigenous tourism is not necessarily a strong and emancipating industrial or societal development factor' (Müller & Viken, 2017, p. 4). A few examples of issues involving tourism are marketing representations (e.g. Niskala & Ridanpää, 2016; Olsen, 2006; Saarinen, 1999), disturbances from tourists (Pettersson & Viken, 2007) and the use of Sámi culture in the tourism industry, especially in Finland (Pettersson, 2004, p. 25). The Sámi are also involved in several local and international political processes to affirm their rights as an indigenous population (http://www.IWGIA.org).

In the process of increased travelling and globalization, local populations can become subordinate to tourism projects, which are often based on a mediated feeling of tradition (Selwyn, 1996). The Sámi have been involved in this process, reconstructing and representing their culture every year during the popular festival in Jokkmokk, Sweden (Müller & Pettersson, 2006). When an indigenous population presents its culture in the context of tourism, the attractive aspects are usually different cultural traits. Tourists search for different cultures and kinds of nature (Müller & Viken, 2017), and they have expectations of seeing such things (Müller & Kuoljok Huuva, 2009; Olsen, 2006; Viken, 2006), which can be influenced by marketing (Costa, 1998; Niskala & Ridanpää, 2016; Schellhorn & Perkins, 2004). In tourism advertising and promotion, the image of indigenous populations and the concept of authenticity are often driven by travel literature, which may be influenced by *otherness*, as is true for the Sámi (Saarinen, 1999). In the end, tourists celebrate otherness (Viken & Müller, 2017).

This paper first shows the conceptualizations of authenticity and heritage relevant for this study. The methodology is then discussed, followed by a presentation and problematization of the analysed data. From the data and the previous research on heritage, Sámi tourism and authenticity, the paper then draws conclusions on the Sámi's representations of their own heritage, and how these are related to more comprehensive discourses on heritage and authenticity. This paper is the result of a constant hermeneutical interpretation (through retroduction and critical realism) between the results of the critical discourse analysis, previous research, and the constantly growing body of research regarding authenticity and the Sámi.

This paper's main goal is to explore a new conceptualization of authenticity based on compromise, through analysing Sámi-produced promotional tourism websites. This is done using a new approach – retroduction and critical discourse analysis. This combination gives an account of the discourses surrounding producing marketing material in the context of tourism.

This paper's novelty is the fact that it compares tourism marketing representations that indigenous people have created to promote themselves to literature that analyses representations produced by third parties. Examples of the latter include: Official websites used by travel agents (Choi, Lehto, & Morrison, 2007; Echtner & Prasad, 2003), printed material for the international market (d'Hauteserre, 2011), or representations to promote a theme park (Magnoni, Ardren, & Hutson, 2007). There are some studies that have already taken into consideration indigenous populations' own creation of promotional material (Fonneland, 2013; Olsen, 2008) or material produced in cooperation with them (Keskitalo & Schilar, 2017). Olsen's study (2008), which is similar to the one proposed in this paper, analyses Maori-produced brochures. This study broadens the scope of this kind of research by taking into consideration several Sámi companies. Furthermore, this paper also presents a new conceptualization of authenticity related to compromise and broadens the scope of the research regarding indigenous tourism. Whitford and Ruhanen (2016, p. 1091) analysed indigenous tourism research and found that indigenous tourism research should be more 'iterative, adaptive and flexible' in style and that stakeholders should participate in the research process. Research should also focus more on the process of creating indigenous tourism and involving indigenous populations. This paper follows these recommendations. The companies whose websites are presented in this research have been contacted by the researcher.

The Sámi and tourism

The Sámi people are usually represented by certain defining cultural markers that are also meant to create difference and entice potential visitors. Some examples include traditional activities, traditional tents or huts, traditional costumes, and reindeer herds (Viken & Müller, 2017). Generally, markers that are used for promoting indigenous populations are related to 'religion, dance, and architecture' (Silver, 1993, p. 304), while the people themselves are represented as primitive. The level of industrialization of the country in which they live is hidden from the promotional messages. Furthermore, cultures are shown as 'timeless and unchanging' (Silver, 1993, p. 304). This is also an important aspect of the 'quest for authenticity' (MacCannell, 1973). This representation gives an image of native populations as just there for tourists' consumption (Silver, 1993). As an example, the Sámi expressed their discomfort over being photographed by tourists, even when they were just going about their daily lives wearing traditional Sámi clothing (Pettersson & Viken, 2007).

Tourism is also a way for the Sámi to re-negotiate their identities (Tuulentie, 2006). They are conscious of their choice to diversify themselves as a way to attract tourists, as can be argued for other indigenous populations (Viken & Müller, 2017). Tourism is also mostly welcome and seen as 'a path to modernity and economic development' (Viken & Müller, 2017, p. 21). Representations related to otherness are also 'challenged and resisted' (Selwyn, 1996, p. 10). As long as the local and indigenous populations have control of their representations and tourism activities involving their cultures, they can represent themselves and their culture as they see fit. However, the development of tourism in Scandinavia is controlled by nonindigenous institutions and DMOs (Destination Marketing Organizations) (Viken & Müller, 2017). The Sámi are also divided about their community's role in tourism, since some see the advantages of commercialization, while others see it as a threat (Pettersson & Viken, 2007).

Cultural commoditization or commercialization is another important issue. One example is whether or not to adapt certain types of cultural expressions (such as the Sámi knife and special cups, called *guksis*) for tourism purposes (Viken & Müller, 2017). The general argument is that commoditization brought by tourism potentially sabotages local cultures. For example, traditional dances and costumes can become commodities for tourists (Cohen, 1988). As Selwyn (1996) argued, the discussion has mainly focused on the fact that commoditization leads to dependency, but this is quite a simplistic way to see things. Cultural exchange has resulted in the creation of other spheres of activity that have demonstrated autonomy, and the festival in Jokkmokk is an example of this (Müller & Pettersson, 2006). Furthermore, commercial development can be a way to do business and sustain the operations of the companies (Viken & Müller, 2017).

Authenticity, heritage and tourism

Heritage tourism is described as a widespread and old type of travel. Heritage is defined as 'present-day use of the past' (Ashworth, 2003; as cited in Timothy & Boyd, 2006, p. 2) and is also connected to political and economic concerns. There is also an issue of authenticity, such as, when tourists visit sites they believe to be authentic and want to have 'authentic encounters' (Timothy & Boyd, 2006). Tourism related to heritage includes travelling to sites where cultures can be observed. Timothy and Boyd (2006) advocated for studying the supply side of heritage tourism, which this paper addresses by looking at indigenous tourism related to Sámi heritage and culture. Indigenous tourism has been defined as 'tourism activities in which indigenous people are directly involved either through control and/or by having their culture serve as the essence of the attraction' (Hinch & Butler, 2006, p. 5). As previously mentioned, tourists often seek authenticity when travelling, which this is also the case for indigenous tourism. This is why it is important to also problematize the supply side of this kind of tourism.

The study of the concept of authenticity is long and diverse. Today, it is mainly about multiple authenticities, rather than only one. There are also many different definitions of authenticity. MacCannell (1973) and Cohen (1988) are two important contributors to this area of research, and Selwyn (1996, p. 6) described the question of authenticity as unresolved, which is still the case nowadays. MacCannell (1973) described it as the reason why tourists travel to some places, while Cohen (1988, p. 374) noted that staged authenticity satisfies tourists' quests for authentic experiences because they search for the 'pristine, the primitive, the natural, that which is as yet untouched by modernity.'

Pearce and Moscardo (1986) proposed an approach to authenticity based on the person and how authenticity could be achieved through different kinds of interactions with places, people or both. Along this same path, Wang (1999) proposed the conceptualization of existential authenticity.

His discussion was based on previous conceptualizations, such as objective (something is either authentic or not) and constructive (authentic is what the visitor perceives as such) authenticities. Existential authenticity is based on 'the existential state of being' that visitors may perceive during touristic activities (Wang, 1999, p. 359).

Authenticity has also been conceptualized as a performance. Therefore, authenticity is neither objective nor subjective, but performative (Zhu, 2012). This type of authenticity is an alternative solution to too objective or too subjective conceptualizations of authenticity. The authenticity of the place is related to the interaction between people, and between people and objects (Knudsen & Waade, 2010). Zhu (2012) argues for a different perspective on performativity; namely that authenticity is related to a connection between heritage and one person's life experiences. To reproduce a ritual for tourism purposes becomes a way to perform identity, as previously mentioned in the case of the Sámi (Tuulentie, 2006). This is a starting point from which to analyse the achievement of authenticity in a particular context (Zhu, 2012). As previously mentioned, promoting tourism can create situations in which the local and indigenous populations are subjects of otherness.

Authenticity is one very influential aspect of this process. Selwyn (1996, p. 28) argues that the myths that are created in the context of tourism are on one hand, characterized by authenticity, while on the other, they are influenced by 'historical, economic and political constructions.' In contrast, Steiner and Reisinger (2006) describe how staging a certain representation for locals may, in fact, be a form of authenticity. The argument that tourism is the sole agent of change, and that the hosts are not able to be agents of change, is in itself, a way to take power from the local population and deprive them from possibly defending themselves (Steiner & Reisinger, 2006).

According to Ram, Björk, and Weidenfeld (2016), there is a connection between tourists' perceptions of a place as authentic and their attachment to it. This also relates to the results of the study by Ivanovic (2014), which empirically shows how history, heritage and 'authentic properties' (p. 512) are positively associated to tourists' experiences. The fact that tourists perceive a place as authentic conveys a sense of value to them. Tourists are not the only ones assigning value to authenticity. Even in the case of UNESCO, there are criteria for authenticity in the requirements for World Heritage Sites (Wang, Huang, & Kim, 2015). Authenticity in this case is the 'cultural value, as well as its value as truthful and credible "information source"' (Wang et al., 2015, p. 1469). One concrete example of this was a complaint in an evaluation protocol by the IUCN (International Union for Conservation of Nature) related to the Sámi using mobile phones on the Sjaunja nature reserve. Use of the mobile phones was seen as damaging for the 'integrity of the environmental sustainability of the area' (Green, 2009, p. 87). As Green argues, a mobile phone is hardly a threat to the environment, but it does not really match with a stereotypical way to see the Sámi and how they should act (2009, p. 87). The studies presented here not only show that authenticity is an important factor for tourists and institutions, but also the harmful effects of the far-reaching authenticity discourse.

Authenticity as compromise

This discussion of relevant literature on authenticity has shown that institutions (such as UNESCO) and tourists make this concept important. It is also clear that there is a very close connection between tourism, authenticity, and heritage. The effects of discourse on authenticity were also introduced. This study contends that previous conceptualizations of authenticity are problematic, since they are based on assumptions that are either too objective or too subjective.

Taylor (2001) has argued for sincerity instead of authenticity, and connects it to performance of culture. He discusses authenticity based on time and the fact that indigenous populations are seen as different because they are also seen as timeless. He argues that realistic views of authenticity are based on criteria spread by marketing. He also quotes MacCannell (1992, pp. 27–35), who describes how combining modernity with traditional aspects of a culture is just another negative aspect of modern economic relations. People are not primitive anymore, but they act as such for economic gain. Taylor (2001, p. 13) asks a very relevant question as to who should 'hold the power to define the authenticity of a cultural experience?' To answer this question – the power should be in the hands of the indigenous populations, for which this new conceptualization of authenticity advocates. Taylor (2001, pp. 11–15) also differs from MacCannell by arguing that even in cases of staged authenticity, those who are performing their culture have an interaction with the tourists that is relevant to their identity, as previously mentioned in the case of the Sámi. Taylor (2001) also mentions that interaction with tourists is a possibility for the Maori to address stereotypes.

In this paper, a new way to conceptualize authenticity based on compromise is presented. However, the term *authenticity* will be used, rather than *sincerity*. The reason is that one suggested solution to the issues related to authenticity could be to instead introduce sincerity (Taylor, 2001). However, it is not possible to fully address this issue in this paper. The main point is not which word is used, but the essence of the terms, which are recognized and searched by both tourists and institutions. In other words, making a division between authenticity and sincerity does not solve the problem of how to define and conceptualize authenticity. In this paper, it is also not possible to address authentication, which has been recently proposed as another possible solution to the above-mentioned debate (Mkono, 2013).

Authenticity as a compromise is based on the work of Wilson (2014), who discusses authenticity from a critical realist viewpoint. Different social structures can affect if we can be authentic to ourselves regarding our ideas and our actions as concrete individuals, such as working in a capitalist setting. Wilson shows how authenticity should be conceptualized in terms of doing, making, and agency. This has been previously argued regarding performance of heritage in the context of tourism (Zhu, 2012). The description of authenticity is discussed using the situation of Early Music musicians' struggle with authenticity. On the one hand, producing Early Music must be historically faithful, but also a personal performance. The solution is a compromise between these two aspects (Wilson, 2014).

A view of authenticity based on compromise allows for principles of authentication to be applied, as well as leaving room for subjective interpretations and a certain performance of one's heritage. The compromise is also expressed in the practical situation of Sámi tourism entrepreneurs, which are often a constant fluctuation and contain conflicts related to their private lives, their belonging to the Sámi community (with its social and political implications), and their role as reindeer herders (Åhrén, 2008; Leu & Müller, 2016; Tuulentie, 2006; Viken, 2006). Furthermore, they are also engaged in conflicts regarding the involvement of their heritage in tourism and varying opinions about it (Pettersson & Viken, 2007).

Methods

This research applied retroduction to the study of representations of indigenous populations. Retroduction is a possible solution to conflicts regarding cooperating with indigenous populations in matters of scientific inquiry and indigenous knowledge (Price, 2005). Retroduction is 'a mode of inference, by which we try to arrive at what is basically characteristic and constitutive of [the] structures' (Danermark, Ekstrom, Jakobsen, & Karlsson, 2002, p. 96). These social structures are involved in the interplay that constitutes social reality at a more abstract level, together with more concrete, everyday manifestations (Fairclough, 2010, pp. 231–232) that are present in every interaction. For instance, tourism marketing is shaped by structures that are more abstract, but become concrete in the production of marketing material and its consumption by tourists.

As critical realism postulates the existence of things behind our perceptions of the world, retroduction is about moving from what is empirically observable to the structures that are not immediately observable (Danermark et al., 2002, p. 96). This usually involves identifying patterns. For instance, certain discourses can be more effective than others in convincing some version of the 'truth' (Ackroyd & Karlsson, 2014, p. 18).

This paper also implies that these representations are based on a certain underlying objectivity, as concrete objects or ideas that are represented (Cashell, 2009). The main assumption is that discourse cannot refer to itself, but must be about something else (Bhaskar, 1993, p. 212; as cited in Cashell, 2009, p. 157).

In this study, critical discourse analysis (CDA) based on retroduction is used to map the relationships between authenticity, tourism marketing, and Sámi cultural heritage. *Heritage* is conceptualized as both tangible and intangible aspects that unite a group. It is also seen as adapting, changing and growing, as well as being connected to the past (Edson, 2004; Waterton, Watson, & Silverman, 2017). The term *critical* comes from the fact that CDA's focus is unequal relationships, at the same time as it strives to provide explanations and solutions to overcome these relationships (Fairclough, 2010, pp. 231–232).

Fairclough's approach stems from the analysis of a social problem and is about finding possible obstacles and solutions to it (2010, pp. 234–239). The general marketing representation of the Sámi and other indigenous populations is problematic (e.g. Amoamo & Thompson, 2010; Niskala & Ridanpää, 2016; Olsen, 2006; Saarinen, 1999; Waitt, 1999), but the Sámi can portray themselves and their culture in the way they deem best. However, they are all individuals with different opinions on tourism and the role of Sámi culture within it. Therefore, it is relevant to analyse the representations that the Sámi themselves have created to discuss the themes that they see as important to attract tourists. Furthermore, studying the underlying mechanisms and structures that make these themes relevant will give us a deeper knowledge into the Sámi's own views on touristic promotional material, considering that Sámi culture is often presented in tourism marketing.

The analysis is based on semiotic material that is analysed under the lenses of different theories. This kind of study implies a certain degree of linguistic analysis to demonstrate the interpretations made from the data by establishing connections to different genres and styles (Zotzmann & O'Regan, 2016). Discourses are different 'ways of representing a part of the world from a particular perspective' (Zotzmann & O'Regan, 2016, p. 116), genres are different interactions (such as create a webpage) and styles are ways of being (such as one's role as an entrepreneur). It is also very important to consider interdiscursivity by connecting the analysis of the text to the analysis of the structures (Zotzmann & O'Regan, 2016).

The empirical material

The research material were the representations of Sámi cultural heritage on the websites of eight Sámi tourism enterprises from Sweden that clearly state in the company description that they are Sámi. The examined sections of the websites were analysed in Swedish and the wordings used were later translated into English in order to investigate them in the original language in which they were written. Websites are sometimes quite comprehensive, so it would exceed the scope of this study to analyse them in their entirety. The chosen sections for analysis were the main page, the potential landing page for potential visitors when visiting the website, and other relevant pages describing offered activities and/or the company, depending on the structure of the website and its content (some websites present most of the information on the main page, while others have very detailed subsections). Pictures were also considered and described to ensure a complete analysis of the website's content. Although the analysis was not quantitative, it sought to be balanced. The companies that were analysed were informed of this research ,and the information was anonymized because not all companies agreed to have their name displayed.

Results

Fairclough's (2001, p. 125) methodology starts by identifying a problem and obstacles to its solution. In this case, the problem comprises the representations that generally dominate tourism marketing, and the obstacle is the wide distribution of these representations. The next step was to analyse the network of practices and the relationship to the particular practice. Retroduction also involves identifying patterns, so the researcher started by reading the text and looking at the pictures on the websites of interest. This revealed patterns that were also present in other studies on tourism marketing. This was not a linear process since the author went back and forth between the websites and the literature to look for similarities and discrepancies. The websites showed clear patterns that were also present in a representative study of brochures produced by members of the Maori population and in many other independent studies regarding the Sámi and other indigenous populations (Olsen, 2008). The most prominent themes that emerged during the analysis were:

- ***Theme 1:*** Connection to nature/harmony with nature/peacefulness (Bruner, 1991; Fennell, 2008; Olsen, 2006; Prebensen, 2007; Waitt, 1997), particularly connection to untouched nature (e.g. Echtner & Prasad, 2003; Pashkevich & Keskitalo, 2017; Prestholdt & Nordbø, 2015).
- ***Theme 2:*** Connection to reindeer (Müller & Pettersson, 2006; Niskala & Ridanpää, 2016; Olsen, 2006; Viken, 2006; Vladimirova, 2011).
- ***Theme 3:*** Connection to the past (Müller & Pettersson, 2001; Niskala & Ridanpää, 2016; Saarinen, 1999; Viken, 2006; Vladimirova, 2011).
- ***Theme 4:*** The use of friendly language (Kalmane, 2010, pp. 77–78; Vahid Dastjerdi, Davoudi-Mobarakeh, & Zare, 2012).
- ***Theme 5:*** *Authenticity* as a noun or adjective (Sharma, Jackson & Inbakaran, 2008; Yeoman, Brass & McMahon-Beattie, 2007).
- ***Theme 6:*** The Sámi costume (Müller & Pettersson, 2001; Niskala & Ridanpää, 2016), Sámi huts or tents (Pettersson, 2002), and traditional singing connected to indigenous populations or the traditional Sámi *yoik* chant (Waitt, 1997, 1999; Bruner, 2001; Pettersson, 2002; Vladimirova, 2011).
- ***Theme 7:*** Joining the Sámi (Amoamo & Thompson, 2010; Keskitalo & Schilar, 2017; Santos, Belhassen, & Caton, 2008).
- ***Theme 8:*** Modernity and useful information for tourists (Müller & Pettersson, 2001; Olsen, 2008; Overend, 2012; Silvén, 2012; Viken & Müller, 2006).

These themes function as an interpretation key for the texts and pictures on the websites. The next phase of Fairclough's (2001) approach is analysing the discourse itself and its relationship to structures, other discourses, and language. Table 1 summarizes these results to simplify the presentation of the data. Each instance identified on the websites was connected to one of the themes and given a general title describing what was presented in either text or picture, or both. These connections were not meant to be neutral, but were the interpretations made by this researcher's particular point of view, mediated by previous research. If the texts mentioned nature in relation to the company or Sámi culture, this phrase was categorized in Theme 1. Since this is not a quantitative study, there is no indication if the reference to a theme was present on a website only once or several times. The only times in which themes were repeated is when one instance is very different from another, such as a mention of both the traditional hut and the costume.

Discussion

The summary of the collected textual and pictorial instances (Table 1) shows that certain themes are more common than others. As Olsen (2008) illustrated, the theme of modernity, which was shown in all of the websites, can be found together with more traditional elements, such as the connection to (untouched) nature, and to peace and harmony, which were also present in all of the websites. The language for addressing tourists is usually friendly, expressed in the use of imperative form ('Follow us!' 'Join us!'), but the capitalization of the word *Ni* (plural form of the personal pronoun *you* in Swedish) could express a form of distancing from tourists and also a form of respect that is reminiscent of some colonial sentiment, since it is not otherwise often used. Only three websites used this form of language.

Three websites explicitly mentioned a connection to nature in their text, but seven of eight company websites contained images related to untouched nature in videos or pictures. As Table 1 shows,

Table 1. Itemization of selected web pages and thematic instances.

Website 1 (Page describing the company)	Use of 'you' plural with capital letter for respect (theme 4) Calmness of the place/Enjoy the silence/phones don't work (theme 1) Reindeer in text and pictures (theme 2) Mention of the hut (theme 6) Traditional costume in pictures (theme 6) Expressions about doing things together (theme 7) Pictures of the family/mention of work outside of herding (theme 8)
Website 2 Page describing the company)	Word authentic (theme 5) Word calm (theme 1) Use of old name used to describe a Sámi settlement (theme 3) Reindeer in text and pictures (theme 2) Mention of modern Sámi cuisine and Sámi institutions (theme 8)
Website 3 (Page describing the company)	Talk to the tourists directly. 'you' singular but capital for respect (theme 4) Reindeer in text and pictures (theme 2) Magnificent nature (theme 1) Cooperation mentioned with other institutions (theme 8) Description of the past of the area (theme 3)
Website 4 (Page describing the company)	Use of 'you' plural with capital letter for respect (theme 4) Magnificent nature/harmony/images of nature with a romantic undertone/ company in the middle of nature (theme 1) Verb used: harmonise (theme 1) Romantic past: people didn't lock their doors (theme 3) Reindeer in text and pictures (theme 2) The yoik mentioned (theme 6) Mention of cooperation/only modernity shown in pictures/ mention of modern Sámi cuisine (theme 8)
Website 5 (Main page – same as company description + activities)	The text talks directly to the reader (theme 4) Campfire association to Sámi culture (theme 5) Activities offered are in peace and harmony with nature (theme 1) Old stove (theme 3) Reindeer in text and pictures (theme 2) Description of the hut present (theme 6) Join the everyday life (theme 7) Separation from the Sámi in the past/no costume (theme 8)
Website 6 (Page about the company)	Verbs to follow with/to discover (theme 4 and theme 7) Connection to nature and ecotourism (theme 1) Reindeer in text and pictures (theme 2) Sámi costume and Sámi hut (theme 6) Talk of Sámi modernity and Sámi institutions (theme 8)
Website 7 (Main page – same as company description + activities main page)	Imperative form such as 'follow us/come with us' (theme 4) Sámi doing things as in the past (theme 3) Reindeer in text and pictures (theme 2) Sámi costume in pictures (theme 6) Picture hinting to join connected to food and fire (theme 7) Mention of Sámi institutions/pictures showing modernity/mention of cooperation/mention of contemporary Sámi activities (theme 8)
Website 8 (Page about the company)	Entrepreneurs refer to themselves as 'we' (theme 4 and theme 7) Authentic objects available there (theme 5) Magnificent land (theme 1) Connection to nature (theme 1) Yoik (theme 6) Join in the everyday life (theme 7) Aspects of modern Sámi life as entrepreneurs (theme 8)

the concept of something untouched is often presented in tourism marketing, regarding both nature and the Sámi (Pettersson, 2004). Untouched nature is meant to attract tourists, but in relation to the Sámi, it is interpreted as a way for them to construct their culture (Valkonen & Valkonen, 2014). Associating the Sámi with untouched nature can also underline the presence of wilderness and the Sámi populations associated with it. Table 1 also includes the category of *being in harmony with nature*, which is another trait that is often associated with the Sámi and other indigenous populations in tourism marketing. The same can be said about the word *peaceful*, which is often associated with both destinations and people in tourism marketing. Expressions related to peace, calm and a connection to nature can be found in almost all of the websites, while more than half of the analysed companies showed a connection to the past. As the literature mentioned in Table 1 shows, the Sámi are often associated to the past. These images are connected to romantic images presented not only in other types of marketing material, but also in historic literature and media representations of the Sámi (Valkonen & Valkonen, 2014).

Reindeer herds were either textually or visually shown in all but one website. The question of associating reindeer with the Sámi is multifaceted. This connection is quite prevalent in tourism marketing in both Nordic countries (Niskala & Ridanpää, 2016) and Russia (Vladimirova, 2011). Furthermore, the Sámi are the only people allowed to herd reindeer in Sweden (Josefsen, 2007). This must be taken into consideration when discussing reindeer herds related to the Swedish Sámi. The reindeer herds are a very important aspect of Sámi culture. They represent a means of sustenance and a symbol that often represents Sámi culture in tourism marketing. Another important aspect to consider is the fact that not all Sámi who are legally allowed to herd reindeer actually choose to do so.

Three of the company websites showed traditional Sámi clothing, three showed Sámi huts or tents, and two mentioned traditional *yoik* singing. The Sámi are often represented in their traditional clothing, in the context of tourism marketing (Table 1). However, there are problems with this, such as tourists taking pictures of the Sámi when they are performing everyday tasks (Pettersson & Viken, 2007). The Sámi hut is also at the centre of tourism marketing for the countries in which the Sámi live. It was also mentioned as part of the activities Sámi entrepreneurs offer (Pettersson, 2002). All the companies also took great effort to show Sámi modernity, such as pictures of the Sámi in everyday clothing or using modern technology, such as snowmobiles and mobile phones. They also describe different cooperations with other actors or institutions. Five websites showed a connection to the Sámi past and historical roots.

The least-used theme involved the word *authenticity* as either an adjective or a noun (only three websites contained the explicit word). There can be multiple retroductive interpretations for this. It may not be necessary to call something authentic because its representation comes from a Sámi company, making it authentic. Another possible interpretation is a conscious refusal to use a word used in marketing (Taylor, 2001). The word *authentic* could also paradoxically draw attention to the possibility that some aspects of the culture might be inauthentic and undermine the reader's trust in the cultural portrayal that is presented. This conceptualization of authenticity will be discussed in future phases of this research.

Conclusions

In the context of this analysis, web pages from eight Sámi tourism companies were analysed to understand how authenticity is structured in the context of tourism marketing and argue for its relationship to a novel conceptualization of authenticity based on compromise.

The last phases of Fairclough's (2001, p. 125) critical discourse analysis framework consider why the social order must address the encountered problem. The instances presented and discussed here are not necessarily problems to be solved or obstacles to overcome, but it is clear from this analysis that there are certain recurring themes that the Sámi use, which are also commonly used elsewhere in tourism. As Viken and Müller (2017) argue, indigenous tourism is not neutral, as its activities are strongly influenced by politics, economy, and ideology. Its language and expressions were analysed and connected to the underlying discourse concerning indigenous populations in tourism marketing. The analysis also shows characteristics that are particularly connected to the situation of indigenous populations in modern times.

This paper argues that modern and traditional aspects of Sámi culture are an expression of an evolving heritage and identity. As Laven (2015, p. 176) contends: 'Heritage development can be a constructive, empowering force for enhancing community-level resilience.' Heritage is not something static that can be preserved. Sámi culture is very much alive. Film festivals, art installations, activism, fashion and music celebrations are only one small aspect of an extremely varied and active cultural Sámi life. Heritage is also adaptive (Edson, 2004) and, similar to authenticity, is connected to agency and doing. The retroductive idea of *compromise* best explains the characteristics of the discourse in tourism websites on Sámi culture.

This paper's analysis of the themes identified in the websites and common themes presented in other marketing websites produced by both indigenous and non-indigenous populations can be interpreted in several ways. Olsen (2008) saw it as an instance of self-orientalism, in which indigenous people present themselves as the colonial powers would. Another interpretation is that the overlying social structures do influence tourism marketing in relation to otherness and authenticity, but that there are also resistance and change driven by the Sámi entrepreneurs.

The tourism discourse identified in marketing is based on a set of binary oppositions between modernity connected to alienation and authenticity (Olsen, 2008). The social structures present in the context of tourism influence the content of these websites, but the themes also show clear connections to Sámi cultural heritage, conceptualized through certain concrete and abstract markers that potential visitors recognize. As previously discussed, representation is grounded in reality, and discourse cannot be about itself. The connection to nature, reindeer herds, and traditional clothing are part of Sámi culture. The representations of these websites all have certain themes in common, so this analysis shows some underlying conceptualization of Sámi culture, represented by certain markers. They show dynamic cultural development, in which there are clear historical connections but also many modern traits. Tourists are looking for difference. This is also why some of the themes described here are needed in the social order (Fairclough, 2001, p. 125). Showing difference is the selling point that makes these companies unique in the general tourism market. There can be a division on the use of culture, mentioned in the previous section, in which some see the advantages of commercialization, while others see it as a threat.

Members of the Sámi population's representation of Sámi heritage and dialogue with the people behind the Sápmi Experience label (International Ecotourism Society, n.d.) should be the basis for national and regional tourism marketing of Sápmi, particularly regarding both national and international agencies and guides. The previously described issues regarding the problems the Sámi have with the presentation of their culture and heritage can only be avoided if they are involved. As Keskitalo (2017, p. 35) noted, there is reason to wonder if 'more accurate points of representation could be found to address this multiplicity of both areas and peoples.' This does not mean that even if the Sámi participate in producing marketing material, the results will be accepted by most of the Sámi population. Rather, including the Sámi in designing new material starts to give them a voice in a process from which they have often been excluded. This is also a way past the obstacles Fairclough (2001, p. 125) mentioned, which were identified in the context of general tourism and indigenous tourism marketing.

Authenticity can be conceptualized as compromise based on the analysed texts because of different aspects. The website as a genre usually contains short bits of text and pictures, making it difficult to balance correct information with attractive messages. Furthermore, without the markers that tourists seek, there may be no visitors to the website. However, all of the companies want to show authentic Sámi heritage by stressing certain links to tradition, but at the same time show that the Sámi have a modern life. This connects to the everyday struggles of Sámi life, including their political struggle, their fight against stereotypes, and tourism as a way to celebrate Sámi heritage (Tuulentie, 2006). The themes also show how Sámi entrepreneurs conceptualize authenticity and why they reject previous conceptualizations. Too much subjectivity can lead to too much dispersion and possible loss of cohesion in the Sámi community. If the conceptualization is too objective, it can lead to conflicts over who gets to decide what is authentic. Authenticity as a compromise allows for adopting several viewpoints, such as performance, and heritage, as well as a certain degree of authentication that provides a more ethical use of Sámi culture in tourism. This perspective on authenticity is a middle ground between historical and cultural accuracy, and individual expression. This compromise is made in light of the underlying discourses surrounding Sámi cultural representations and tourism in general. The label *Sápmi Experience* (http://www.fjallhasten.se/sapmi-experience/) created and controlled by members of the Sámi population, is an expression of this compromise.

The last phase of Fairclough's (2001, p. 125) framework is a general reflection on the whole analysis. This study has attempted to achieve a well-rounded discussion of the marketing themes that are present on Sámi websites. Pictures, texts, and general discourses were analysed to show that there are structures that underlie all tourism marketing, both Sámi and non-Sámi. The descriptions that Sámi entrepreneurs publish on their websites can be connected to a characterization of authenticity as a compromise, in which the Sámi must balance several aspects of their lives. As tourism entrepreneurs, they want to attract visitors and show all the interesting aspects of Sámi culture. However, that is not easy, considering the expectations that marketing discourses can entice in potential customers. In addressing this issue, this study also has limitations. It is missing the entrepreneurs' and local indigenous populations' point of view about what is important to display for marketing purposes. This should be addressed by future research, which is planned for this research project.

Disclosure statement

No potential conflict of interest was reported by the author.

ORCID

Cecilia de Bernardi http://orcid.org/0000-0002-1400-0357

References

Ackroyd, S., & Karlsson, J. C. (2014). Critical realism, research techniques, and research designs. In P. Edwards, J. O'Mahoney, & S. Vincent (Eds.), *Studying organizations using critical realism: A practical guide* (pp. 21–45). New York: Oxford University Press.

Amoamo, M., & Thompson, A. (2010). (Re)Imaging māori tourism: Representation and cultural hybridity in postcolonial New Zealand. *Tourist Studies*, *10*(1), 35–55.

Åhrén, C. (2008). *Är jag en riktig same? En etnologisk studie av unga samers identitetsarbete*. Umeå: Umeå University.

Bruner, E. (1991). Transformation of self in tourism. *Annals of Tourism Research*, *18*(2), 238–250.

Bruner, E. (2001). The maasai and the lion king: Authenticity, nationalism, and globalization in African tourism. *American Ethnologist*, *28*(4), 881–908.

Cashell, K. (2009). Reality, representation and the aesthetic fallacy: Critical realism and the philosophy of C. S. Peirce. *Journal of Critical Realism*, *8*(2), 135–171.

Choi, S., Lehto, X. Y., & Morrison, A. M. (2007). Destination image representation on the web: Content analysis of Macau travel related websites. *Tourism Management*, *28*(1), 118–129.

Cohen, E. (1988). Authenticity and commoditization in tourism. *Annals of Tourism Research*, *15*(3), 371–386.

Costa, J. A. (1998). Paradisal discourse: A critical analysis of marketing and consuming hawaii. *Consumption Markets & Culture*, *1*(4), 303–346.

Danermark, B., Ekstrom, M., Jakobsen, L., & Karlsson, J. C. (2002). *Explaining society: Critical realism in the social sciences*. London: Routledge.

d'Hauteserre, A.-M. (2011). Politics of imaging New Caledonia. *Annals of Tourism Research*, *38*(2), 380–402.

Echtner, C. M., & Prasad, P. (2003). The context of third world tourism marketing. *Annals of Tourism Research*, *30*(3), 660–682.

Edson, G. (2004). Heritage: Pride or passion, product or service? *International Journal of Heritage Studies, 10*(10), 333–348.
Fairclough, N. (2001). Critical discourse analysis as a method in social scientific research. In R. Wodak, & M. Meyer (Eds.), *Methods of critical discourse analysis* (pp. 122–138). London: Sage.
Fairclough, N. (2010). *Critical discourse analysis: The critical study of language*. London: Pearson Education.
Fennell, D. (2008). Ecotourism and the myth of indigenous stewardship. *Journal of Sustainable Tourism, 16*(2), 129–149.
Fonneland, T. A. (2013). Sami tourism and the signposting of spirituality: The case of Sami tour: A spiritual entrepreneur in the contemporary experience economy. *Acta Borealia, 30*(2), 190–208.
Green, C. (2009). *Managing laponia: A world heritage site as arena for Sami ethno-politics in Sweden (Doctoral dissertation)*. Uppsala University, Uppsala.
Hinch, T., & Butler, R. (2006). Introduction: Revisiting common ground. In R. W. Butler & T. Hinch (Eds.), *Tourism and indigenous peoples* (pp. 1–12). Amsterdam: Elsevier.
International Ecotourism Society. (n.d.). *Sápmi experience label recognizes excellence in authentic and sustainable tourism*. Retrieved from http://www.ecotourism.org/news/sápmi-experience-label-recognizes-excellence-authentic-and-sustainable-tourism
Ivanovic, M. (2014). The perceived authenticity of iconic heritage sites in urban tourism: The case of constitutional hill, Johannesburg, South Africa. *Urban Forum, 25*(4), 501–515.
Josefsen, E. (2007). The saami and the national parliaments: Channels for political influence. Gáldu Čála. *Journal of Indigenous Peoples Rights, 2*. Retrieved from http://archive.ipu.org/splz-e/chiapas10/saami.pdf
Kalmane, M. R. (2010). *Advertising: Using words as tools for selling*. Lulu Enterprises UK Ltd.
Keskitalo, E. C. H. (2017). Images of the northern and 'Arctic' in tourism and regional literature. In A. Viken & D. K. Müller (Eds.), *Tourism and indigeneiry in the Arctic* (pp. 36–49). Bristol: Channel View.
Keskitalo, E. C. H., & Schilar, H. (2017). Co-constructing 'northern' tourism representations among tourism companies, DMOs and tourists. An example from jukkasjärvi, Sweden. *Scandinavian Journal of Hospitality and Tourism, 17*(4), 406–422.
Knudsen, B. T., & Waade, A. M. (2010). Performative authenticity in tourism and spatial experience: Rethinking the relation between travel, place and emotion in the context of cultural economy and emotional geography. In B. T. Knudsen & A. M. Waade (Eds.), *Re-investing authenticity* (pp. 1–19). Bristol: Channel View.
Laven, D. (2015). Heritage development and community resilience: Insights for the era of climate change. In D. C. Harvey & J. Perry (Eds.), *The future of heritage as climate change – loss, adaptation, and creativity* (pp. 167–179). New York: Routledge.
Leu, T. C., & Müller, D. K. (2016). Maintaining inherited occupations in changing times: The role of tourism among reindeer herders in northern Sweden. *Polar Geography, 39*(1), 40–57.
MacCannell, D. (1973). Staged authenticity: Arrangements of social space in tourist settings. *American Journal of Sociology, 79*(3), 589–603.
MacCannell, D. (1992). *Empty meeting grounds [elektronisk resurs]*. London: Routledge.
Magnoni, A., Ardren, T., & Hutson, S. (2007). Tourism in the mundo maya: Inventions and (mis)representations of maya identities and heritage. *Archaeologies, 3*(3), 353–383.
Mkono, M. (2013). Hot and cool authentication: A netnographic illustration. *Annals of Tourism Research, 41*, 215–218.
Müller, D. K., & Kuoljok Huuva, S. (2009). Limits to Sami tourism development: The case of Jokkmokk, Sweden. *Journal of Ecotourism, 8*(2), 115–127.
Müller, D. K., & Pettersson, R. (2001). Access to Sami tourism in northern Sweden. *Scandinavian Journal of Hospitality and Tourism, 1*(1), 5–18.
Müller, D. K., & Pettersson, R. (2006). Sámi heritage at the winter festival in jokkmokk, Sweden. *Scandinavian Journal of Hospitality and Tourism, 6*(1), 54–69.
Müller, D. K., & Viken, A. (2017). Indigenous tourism in the Arctic. In A. Viken & D. K. Müller (Eds.), *Tourism and indigeneity in the Arctic* (pp. 3–32). Bristol: Channel View.
Niskala, M., & Ridanpää, J. (2016). Ethnic representations and social exclusion: Sáminess in Finnish lapland tourism promotion. *Scandinavian Journal of Hospitality and Tourism, 16*(4), 375–394. doi:10.1080/15022250.2015.1108862.
Olsen, K. (2006). Making differences in a changing world: The Norwegian Sámi in the tourist industry. *Scandinavian Journal of Hospitality and Tourism, 6*(1), 37–53.
Olsen, K. (2008). The maori of tourist brochures representing indigenousness. *Journal of Tourism and Cultural Change, 6*(3), 161–184.
Overend, D. (2012). Performing sites: Illusion and authenticity in the spatial stories of the guided tour. *Scandinavian Journal of Hospitality and Tourism, 12*(1), 44–54.
Pashkevich, A., & Keskitalo, E. C. H. (2017). Representations and uses of indigenous areas in tourism experiences in the Russian Arctic. *Polar Geography, 40*(2), 85–101.
Pearce, P. L., & Moscardo, G. M. (1986). The concept of authenticity in tourist experiences. *The Australian and New Zealand Journal of Sociology, 22*(1), 121–132.

Pettersson, R. (2002). Sami tourism in northern Sweden: Measuring tourists' opinions using stated-preference methodology. *Tourism and Hospitality Research, 3*(4), 357–369.

Pettersson, R. (2004). *Sami tourism in northern Sweden : Supply, demand and interaction (Doctoral dissertation)*. Umeå University, Sweden.

Pettersson, R., & Viken, A. (2007). Sami perspectives on indigenous tourism in Northern Europe: Commerce or cultural development? In R. W. Butler & T. Hinch (Eds.), *Tourism and indigenous peoples* (pp. 176–187). Amsterdam: Elsevier.

Prebensen, N. K. (2007). Exploring tourists' images of a distant destination. *Tourism Management, 28*(3), 747–756.

Prestholdt, R., & Nordbø, I. (2015). Norwegian landscapes: An assessment of the aesthetical visual dimensions of some rural destinations in Norway. *Scandinavian Journal of Hospitality and Tourism, 15*(1–2), 202–222.

Price, L. (2005). Playing musement games: Retroduction in social research, with particular reference to indigenous knowledge in environmental and health education. *South African Journal of Environmental Education, 22*, 87–96.

Ram, Y., Björk, P., & Weidenfeld, A. (2016). Authenticity and place attachment of major visitor attractions. *Tourism Management, 52*, 110–122.

Saarinen, J. (1999). Representation of indigeneity: Sami culture in the discourses of tourism. In J. N. Brown & P. M. Sant (Eds.), *Indigeneity: Construction and Re/presentation* (pp. 231–249). Commack, NY: Nova Science.

Santos, C. A., Belhassen, Y., & Caton, K. (2008). Reimagining chinatown: An analysis of tourism discourse. *Tourism Management, 29*(5), 1002–1012.

Schellhorn, M., & Perkins, H. C. (2004). The stuff of which dreams are made: Representations of the south Sea in German-language tourist brochures. *Current Issues in Tourism, 7*(2), 95–133.

Selwyn, T. (1996). Introduction. In T. Selwyn (Ed.), *The tourist image: Myths and myth-making in tourism* (pp. 1–32). Chichester: John Wiley and Sons.

Sharma, M., Jackson, M., & Inbakaran, R. (2008). Promoting indigenous food to foreign visitors: An Australian study. *Asian Anthropology, 7*(1), 121–134.

Silver, I. (1993). Marketing authenticity in third world countries. *Annals of Tourism Research, 20*(2), 302–318.

Silvén, E. (2012). *Contested sami heritage : Drums and sieidis on the move. In: National museums and the negotiation of difficult pasts conference proceedings from EuNaMuS, identity politics, the uses of the past and the European citizen (pp. 173–186).* Linköping: Linköping University Electronic Press.

Steiner, C. J., & Reisinger, Y. (2006). Understanding existential authenticity. *Annals of Tourism Research, 33*(2), 299–318.

Taylor, J. P. (2001). Authenticity and sincerity in tourism. *Annals of Tourism Research, 28*(1), 7–26.

Timothy, D. J., & Boyd, S. W. (2006). Heritage tourism in the 21st century: Valued traditions and new perspectives. *Journal of Heritage Tourism, 1*(1), 1–16.

Tuulentie, S. (2006). The dialectic of identities in the field of tourism. The discourses of the indigenous sámi in defining their own and the tourists'. Identities. *Scandinavian Journal of Hospitality and Tourism, 6*(1), 25–36.

Vahid Dastjerdi, H., Davoudi-Mobarakeh, S., & Zare, M. (2012). A multimodal analysis of differences between TV commercials and press advertisements: A discoursal study of persuasion-seeking strategies in the mass media. *Studies in Literature and Language, 4*(2), 115–121.

Valkonen, J., & Valkonen, S. (2014). Contesting the nature relations of sami culture. *Acta Borealia, 31*(1), 25–40.

Viken, A. (2006). Tourism and sámi identity – An analysis of the tourism-identity nexus in a sámi community. *Scandinavian Journal of Hospitality and Tourism, 6*(1), 7–24.

Viken, A., & Müller, D. K. (2006). Introduction: Tourism and the sámi. *Scandinavian Journal of Hospitality and Tourism, 6*(1), 1–6.

Viken, A., & Müller, D. K. (2017). Indigeneity and indigenous tourism. In A. Viken & D. K. Müller (Eds.), *Tourism and indigeneity in the Arctic* (pp. 16–32). Bristol: Channel View Publications.

Vladimirova, V. K. (2011). "We are Reindeer People, We Come from Reindeer." Reindeer herding in representations of the Sami in Russia. *Acta Borealia, 28*(1), 89–113.

Waitt, G. (1997). Selling paradise and adventure: Representations of landscape in the tourist advertising. *Australian Geographical Studies, 35*(1), 47–60.

Waitt, G. (1999). Naturalizing the 'primitive': A critique of marketing Australia's indigenous peoples as 'hunter-gatherers.' *Tourism Geographies: An International Journal of Tourism Space, Place and Environment, 1*(2), 142–163.

Wang, N. (1999). Rethinking authenticity in tourism experience. *Annals of Tourism Research, 26*(2), 349–370.

Wang, Y., Huang, S., & Kim, A. K. (2015). Toward a framework integrating authenticity and integrity in heritage tourism. *Journal of Sustainable Tourism, 23*(10), 1468–1481.

Waterton, E., Watson, S., & Silverman, H. (2017). An introduction to heritage in action. In H. Silverman, E. Waterton, & S. Watson (Eds.), *Heritage in action: Making the past in the present* (pp. 3–16). Cham, Switzerland: Springer.

Whitford, M., & Ruhanen, L. (2016). Indigenous tourism research, past and present: Where to from here? *Journal of Sustainable Tourism, 24*(8–9), 1080–1099.

Wilson, N. (2014). Managing authenticity. *Journal of Critical Realism, 13*(3), 286–303.

Yeoman, I., Brass, D., & McMahon-Beattie, U. (2007). Current issue in tourism: The authentic tourist. *Tourism Management*, *28*(4), 1128–1138.

Zhu, Y. (2012). Performing heritage: Rethinking authenticity in tourism. *Annals of Tourism Research*, *39*(3), 1495–1513.

Zotzmann, K., & O'Regan, J. P. (2016). Critical discourse analysis and identity. In S. Preece (Ed.), *The routledge handbook of language and identity* (pp. 113–128). London: Routledge.

Indigenous tourism in Australia: understanding the link between cultural heritage and intention to participate using the means-end chain theory

Trinidad Espinosa Abascal

ABSTRACT

Tourism is often suggested as a development strategy for Indigenous people while protecting their cultural and natural heritage. However, without actual visitor participation, the good intentions of tourism cannot be achieved. By using the means-end chain theory and adopting a convergent mixed-methods approach using photographs, this study seeks to identify the link between visitor perceptions about Australian Indigenous tourism attributes and their intention to participate on this type of tourism. Two Australian Indigenous tourism activities (short tour and performance), portraying both traditional and contemporary culture, at Cairns, Queensland were tested with 209 visitors. The results reveal that the attributes 'history/art/culture' and 'local' are perceptions linked to Indigenous tourism. The findings support previous studies by identifying that some participants associated Indigenous tourism with a traditional past; however, the majority of participants expressed a contemporary perception of Indigenous culture. These findings highlight the need for Indigenous tourism stakeholders to monitor and protect their historical and evolving cultural heritage (in particular intangible heritage) when engaging in Indigenous tourism. Cultural heritage could be susceptible to commodification and loss of authenticity in order to accommodate visitor expectations.

Introduction

It has been claimed that Indigenous tourism is as a strategy to increase economic benefit, to revive the local Indigenous culture and to help maintain cultural heritage (Australian Government, 2015; Dyer, Aberdeen, & Schuler, 2003). On the visitors' level, it could increase the level of education about Indigenous culture and enhance their connection with history and land (Abascal, Fluker, & Jiang, 2016). Other Indigenous tourism benefits may include the strengthening of national identity and cross-cultural understanding (Dyer et al., 2003; Galliford, 2009; Higgins-Desbiolles, 2003). However, critics of Indigenous tourism suggest that this type of tourism is based on the 'stereotypical image of a marginalised people' (Ryan & Aicken, 2005, p. 4). They argue that it could not only bring change and/or damage in social values, culture, and habitat within the host community, but also negative consequences such as stress (Hodgson, Firth, & Presbury, 2007), racism, exploitation, assimilation (Ryan, 2005) and the risk of gentrification (Chan, Iankova, Zhang, McDonald, & Qi, 2016). It is also claimed that Indigenous tourism could affect the commodification of Indigenous peoples and their culture; and their rights to self-determination (Whitford & Ruhanen, 2016).

Within Australia, several tourism marketing campaigns have emphasised Australia's Indigenous tourism as a key point of differentiation from other destinations around the world (Pomering & White, 2011) and as a strategy to conserve Indigenous cultural heritage (Australian Government, 2015). However, the consecutive years of declining visitor demand and the rapid growth of the Asian market, in particular China (Ruhanen, Whitford, & McLennan, 2015a; Tourism Research Australia, 2016), raises the question of whether Indigenous tourism is a successful strategy to bring benefits to Indigenous communities. Therefore, to understand this issue, it is important to understand Indigenous tourism consumer demand.

Existing studies on Indigenous tourism from the perspective of consumer demand have addressed issues such as the demographic characteristics of Indigenous visitors, their preferences, awareness, motivations, constraints, and the mental processes they go through when deciding to participate, or not, in Indigenous tourism (Abascal et al., 2016; Abascal, Fluker, & Jiang, 2015; Jones Donald Strategic Partners, 2009; Ruhanen et al., 2015a; Ruhanen, Whitford, & McLennan, 2015b; Ryan & Huyton, 2000, 2002). While previous studies have contributed to the tourism consumer behaviour knowledge, it appears that they have investigated these concepts independently. In addition, there are very few known studies that have focused on understanding tourism activity choices. It has been claimed that a decision to engage in a tourism activity is a trade-off between the perceived benefits (motivations) and the effort required to diminish constraints associated with the activity (Tian, Crompton, & Witt, 1996). Therefore, this study aims to use the means-end theory to investigate the link between visitor intention to participate on two Indigenous tourism activities while visiting a particular destination and visitor perceptions of Indigenous cultural heritage portrayed (traditional and/or contemporary) in the activities. The general aim of this study may be communicated more specifically within the following three explicit research objectives: (1) To explore the visitor intention to participate in two Indigenous tourism activities in comparison with other types of tourism activities offered at the destination; (2) to identify the perceived attributes of Indigenous cultural heritage portrayed in two Indigenous tourism activities; and (3) to explore the link (either as perceived motivations or constraints) between visitor intention to participate and perceived attributes of Indigenous cultural heritage portrayed in two Indigenous tourism activities. As such, this study extends the empirical enquiry to the field by using a mixed-methods approach in a case study to gain insights into these links. This information could then inform strategies to increase demand for this type of tourism in a way that the cultural heritage is properly portrayed and managed within the sector.

Theories on motivations and constraints

For this study, the 'intention to participate' is defined as the visitor's perceived likelihood of participating in a tourism activity within a specific time period – while they were travelling in Cairns – (Ruhanen et al., 2015b). Haas, Driver, and Brown (1980) concluded that when choosing activities, visitors evaluate each alternative as being either positive or negative in comparison with their own personal motivation force. Therefore, it is clear that understanding beliefs regarding motivations and constraints is fundamental to understand intention to participate in tourism activities. Constraints are the opposite of motivations and are defined as 'any factor which intervenes between the preference for an activity and participation in it' (Crawford & Godbey, 1987, p. 120). Crawford and Godbey (1987) suggested that constraints should be classified in three categories: (1) *intrapersonal*, 'involve individual psychological states and attributes which interact with leisure preferences rather than intervening between preferences and participation'; (2) *interpersonal*, 'are the result of interpersonal interaction or the relationship between individuals' characteristics'; and (3) *structural*, are defined as 'intervening factors between leisure preference and participation' (Crawford & Godbey, 1987, pp. 123–124). The development of the constraints theory within the tourism field remains limited. Nevertheless, examples of tourism studies investigating constraints following the intrapersonal–interpersonal-structural typology are extensive. This study only reports interpersonal constraints as they are related to individual perceptions and preferences (rather than intervening factors between preference and actual participation).

During the last few decades, different theories or models of travel motivation (i.e. allocentrics-psychocentric, optimal arousal theory, travel career ladder theory; push–pull theory) have contributed to tourism research (Jiang, Scott, & Ding, 2015). While the majority of these theories focus only on motivations, the push–pull theory suggests that visitors are pushed to travel by invisible factors (internally generated drives – motivations) and pulled by visible factors (attributes, features). This theory has been widely used by researchers. Indeed, there are some studies that have demonstrated the interrelationships between the pull (attributes of a destination) and the push factors (motivations) on travel decisions (Klenosky, 2002).

It has been suggested that the means-end chain theory and the resulting Hierarchical Value Map [HVM] is a good approach to understand visitors' motivations and to examine the push–pull relationships (Jiang et al., 2015; Klenosky, 2002). The means-end theory describes the hierarchical relationships between product/service attributes (the means), the consequences for the consumer provided with these attributes (consequences/benefits) and the personal values (the ends) these consequences reinforce (Gutman, 1982). This theory is broadly used in marketing research to understand perchance and consumption decisions (Klenosky, 2002). In tourism, the means-end theory has been used to explore the push (attributes) and pull (motivational forces important to the visitor expressed as consequences/benefits and personal values) factors regarding destination choice (Jewell & Crotts, 2001; Jiang et al., 2015; Klenosky, 2002). For the present study, the employment of the means-end chain theory allowed the uncovering of linkages between attributes, consequences/benefits and values regarding the intention to participate in Indigenous tourism activities.

Preferences for, and perceptions of, Indigenous tourism

It is often argued that Indigenous tourism is not considered as being a top priority tourism activity which either international or domestic visitors choose to participate in while travelling in Australia (Abascal et al., 2015; Ruhanen et al., 2015a, 2015b; Ryan & Huyton, 2000, 2002). Ryan and Huyton (2000, 2002) suggested that the visitors interested in Indigenous tourism are a small niche market, as for the greater majority of visitors the interest is in the territory as a whole. More recently, Whitford and Ruhanen (2014) suggested that some marketing strategies have had a positive impact on the amount of participation in Australian Indigenous tourism, as certain sites or activities have been established as 'must see' attractions.

Interestingly, despite the low domestic participation rate, domestic visitors are more aware of the Indigenous tourism activities available in Australia compared to international visitors (Ruhanen et al., 2015b). This lack of awareness by the international market is confirmed by Ashwell (2014) and Galliford (2009) who claimed that international visitors associate Indigenous tourism more closely with historical or stereotypical themes (a perception of 'stone age people'), than with the varied contemporary Indigenous tourism activities as they have a limited awareness about this diversity. Domestic visitors have a more realistic understanding of Indigenous people as a current cultural group (Galliford, 2009). However, domestic visitors still believe that by participating in Indigenous tourism they will get connected with a traditional past (Abascal et al., 2016). It has been reported that visitor expectations are built on the romantic and nostalgic (primitive, pristine and unmodernised) concept of 'authenticity' (Mkono, 2016; Pomering & White, 2011). Destination images play an important role on creating these expectations. A recent study conducted by Seiver and Matthews (2016) suggested that the New South Wales destination image lacks reference to Indigenous culture and history, and the little references made tend to be 'static, stereotypical and homogenous' (Seiver & Matthews, 2016, p. 1310). They claimed that when marketing images represent Indigenous people as a living culture worthy of attention, it fosters an attitude of appreciation that could disrupt negative stereotypes (Seiver & Matthews, 2016).

Indigenous tourism perceived attributes, motivations and constraints

Features of Indigenous tourism attractiveness have been identified in different contexts. It appears that 'learning,' 'history,' 'traditional lifestyles,' 'contemporary lifestyle,' 'contact with Indigenous people,' 'natural scenery' and 'authenticity' are the most important attributes (Chang, Wall, & Chu, 2006; Kutzner, Wright, & Stark, 2009; McIntosh & Johnson, 2005; Notzke, 2004; Ryan & Huyton, 2000; Zeppel, 2002). Ingram (2005) conducted a phenomenological study with 17 visitors who visited Alice Springs (and the 'must see' attraction: Uluru). Her findings suggested that positive perceptions include: the power and spirituality of Uluru, appreciation for Indigenous people's survival skills, and interest for various aspects of their culture and meaningful interaction with Indigenous people. Other motivations to participate in Indigenous tourism include: 'discovery and learning,', 'connection with the land,' 'understanding of Indigenous culture,' 'challenge and adventure,' 'fun and enjoyment,' 'reward and satisfaction,' 'learning opportunities for children' and 'appreciation' (Abascal et al., 2015; Jones Donald Strategic Partners, 2009; Ruhanen et al., 2015b). Constraints to participate in Indigenous tourism include: lack of Indigenous tour guides, low quality of cultural information provided by non-Indigenous tourism guides, negative perceptions regarding authenticity and overcommercialisation, preconceptions regarding Indigenous people, feelings that some aspects of Indigenous culture are missing, lack of uniqueness and boring and passive activities (Abascal et al., 2015; Ingram, 2005; Ruhanen et al., 2015b). However, it has been claimed that beliefs regarding 'uniqueness,' 'authenticity,' and 'nature-based environment' could be either a positive or negative influence for intentional behaviour (Abascal et al., 2016).

Previous findings suggested that interest, and intention to participate, in Indigenous tourism depends upon the perceived attribute of being 'authentic'; therefore, acculturation or inauthentic representation could have a negative impact on it (Butler & Hinch, 2007; Dyer et al., 2003). Abascal et al. (2016) and McIntosh and Johnson's (2005) findings suggested that the perception of 'authenticity' also relates to aspects of the setting. Contrary to this, Ryan and Huyton (2002) claimed that the majority of visitors are not really looking for authenticity; the danger is, however, in promoting 'authentic' Indigenous tourism in which visitors believe that they will gain insights into a complex culture in a small amount of time or unreal setting. In fact, Muller and Pettersson (2005) suggested that through Indigenous tourism, visitors usually experience a staged authenticity; nonetheless, this does not necessarily alter their perception of authenticity. McIntosh and Johnson (2005) also added that the 'sincerity' between hosts and visitors, rather than historical accuracy is important for authenticity. However, it is important to point out that authenticity is an evolving concept based on individual perceptions, or beliefs, of reality (McIntosh & Johnson, 2005; Notzke, 2004). In fact, authenticity should be assumed in terms of context and it can never be complete as individuals and communities are complex individuals that interact with others only partially and contextually (Bignall, 2014). Muller and Pettersson (2005, p. 203) stated that in Indigenous tourism, 'Indigenous heritage is constantly objected to negotiations, both within the indigenous groups, and in a wider society'; illustrating that heritage is dynamic. Indeed, it has been suggested that Indigenous tourism promotion needs to include the contemporary and changing reality of many, and diverse, Indigenous cultures (Amoamo & Thompson, 2010).

The study areas: Cairns, QLD

Cairns is located within the Tropical North Queensland region about 1701 kilometres northwest of Brisbane, the capital city of the state of Queensland. Cairns is a major city and is widely known for its accessibility to the Great Barrier Reef (one of the seven natural wonders of the world) and the Daintree Rainforest (which is included in the UNESCO World Heritage Site list). Within Cairns and surrounds, there are three Indigenous countries. One belongs to the Yirriganydji people and their territory is between what is now named as Cairns and Port Douglas. The second country is the Djabuganjdji (also known as Djabugay or Tjapukai). This territory is between Yirriganydji and Kuku

Yalanji's territory. The third group is the Kuku Yalanji people who are known as the rainforest people, and their country extends from Mossman to Cooktown in the north and Palmer River in the west (http://www.mossmangorge.com.au/). Native title was granted to the Djabugay people (17 December 2004) and the Kuku Yalanji people (9 December 2007) (http://www.nntt.gov.au/). The Yirriganydji people submitted a native title claim on the 29 October 2012; however, there is no resolution as yet (*Yirrganydji (irukandji) people v. Queensland* 2015).

Methodology

As previously mentioned, this study aims to explore the link between visitor intention to participate on two Indigenous tourism activities while visiting a particular destination and visitor perceptions of the Indigenous cultural heritage portrayed (traditional and/or contemporary) in the activities. The means-end theory was used to uncover participants' perceived attributes of the Indigenous tourism activities, and to link them with the visitor intention to participate (considering the perceived consequences/benefits, values and constraints associated with the activity). Means-end analysis comprises an interviewing technique known as the laddering technique which consists of the following steps: (1) to identify attributes: participants are asked to mention the attributes of products or services that are important to them; (2) to identify consequences/benefits: participants are asked to describe the perceived benefits associated with these attributes; and (3) to identify personal values: participants are asked why these consequences/benefits are important to them (Walker & Moscardo, 2016).

This study adopted a convergent mixed-methods approach using the ranking-sorting photo-based procedure and semi-structured interviews in a case study (the method has been described in detail in Abascal, Fluker, and Jiang (2018) and will be summarised below). This design involves the contemporaneous data collection and analysis of both quantitative and qualitative data. The focus is on comparing and relating both components of the project with the aim of seeking completeness (Creswell & Plano Clark, 2011). In the present study, this approach was chosen first, because it allows the researcher to enhance and enrich the meaning of a single method (in particular, it enriches the meaning of the quantitative data collected to measure the intention to participate), and second to develop a more complete understanding of the problem – to uncover perceived attributes of the Indigenous tourism activities, consequences/benefits, values and constraints (complementarity).

Validation process

Part of the data collection process involved the use of 10 photographs depicting site-specific tourism activities. To select valid, and easy to interpret, photographic representations of each of the tourism activities under investigation, a validation process was conducted with a convenience sample of ten participants, each of whom had a level of expertise in the broad subject area of this study. These photographs were selected and acquired from the Tourism and Events Queensland website, which categorised them into 'arts & culture,' 'food & wine,' 'Indigenous,' 'outdoor/adventure,' and 'outdoor/nature.' The main tourism activities portrayed in marketing campaigns by the destination were included along with the Indigenous tourism activities. During the validation process participants were asked to look at 50 hard copy printed photographs (4 × 6) of tourism activities that are available at Cairns and the surrounding region and were asked to sort them under each category heading. Participants then were asked to give a name for each activity represented in the photograph (for example: Indigenous short tour). Then the repertory grid technique was deployed. The repertory grid involved participants sorting three elements (in this case photographs), which constitute a triad, and to then specify why two of them are more alike (to obtain constructs) and different from the third (to obtain contrast) (Coshall, 2000). To choose the two activities to represent each tourism category, the following were taken into consideration: (1) the activities with the highest frequency

Table 1. Description of the two Indigenous tourism activities under study.

Image	Identified constructs	Description
 Indigenous short-tour Source: Adventure North Australia	- Nature-based - Outdoor - Food related - Guided learning - Interactive - No traditional painting/dress - Adventure	The main Indigenous tours in the area are (1) guided walks at Mossman Gorge, (2) day experiences at Tjapukai Aboriginal Cultural Park, (3) Walkabout, (4) the Bama-way Daintree dreaming tour, and (5) Pamagirri Aboriginal Experience within RainForeStation. Within those tours there is a mix between traditional and contemporary portrayals of Indigenous culture. These include: traditional smoking ceremony, stories of culture and traditions, visitation to ceremony sites, paint demonstration, hunting and weapons demonstration, didgeridoo show, boomerang and spear throwing, traditional fishing and gathering techniques, and bush tucker.
 Indigenous performance Source: Tourism and Events Queensland	- Staged - No interaction - Traditional painting/dress - Spectacle-based	Visitors can experience a night at 'Tjapukai Aboriginal Cultural Park' or at 'Flames of the Forest'. Both include elements of music, dance and traditional painting. They also use technology in forms of light and sound. In addition, during the data collection stage, the Cairns Indigenous Art Fair [CIAF] was on, so visitors could experience Aboriginal and Torres Islander performances as part of it.

within the correspondent tourism category; (2) the activity name given by participants with the highest frequency; and (3) the constructs identified for each activity.

The two Indigenous tourism activities chosen during the validation process were the 'short tour' and the 'performance'. To avoid participants' own interpretation of the photographs, a small description of each photo was included above each image. The selected photographs used during the data collection process to depict these activities are shown in Table 1. The information on the table suggests that the two Indigenous tourism activities under study portray both traditional and contemporary Indigenous cultural elements.

Data collection process

The in-field data collection process was held from the 27 July to the 13 August 2015 with both domestic and international visitors within the Visitor Information Centre at Cairns, Queensland. A convenience sample of visitors entering the Visitor Information Centre was invited to participate in the study using the mall intercept technique (Butler, 2008). A total of 209 complete procedures were conducted. The main reasons given for not wanting to participate in the survey were: not enough time (46), not interested (36), language barriers (10), and other (5). During the data collection process, first the ranking-sorting procedure was deployed. A semi-structured interview followed by using the means-end chain theory and the associated laddering technique by asking participants to explain the reasons for their intention to participate, or not, in the ten tourism activities. The interview process was terminated when either no new information was being obtained, or the participant chose to finish the interview. After that, participants were asked to fill in a survey to collect demographics, psychographics and travel behaviour data. Each procedure took approximately 15–25 min. Each interview was audio-recorded and hand-written notes were taken by the researcher. Transcriptions of the interviews were analysed using content analysis. The data analysis process started first by classifying the responses into either: (a) when participants intended to participate; or (b) participants did not intend to participate. Second, categories were generated following different frameworks: the coding of the attributes was conducted by an inductive approach (from the data). The coding at the consequences/benefits level followed the pool of motivations used in Pearce and Lee's (2005) study. The coding at the value level followed the literature on human values (Kahle, 1983; Rokeach, 1973). To code the lack of intention to participate, the constraints typology of Crawford and Godbey (1987) was used. However, as previously mentioned, this study only reports intrapersonal constraints related to the perception of the activities' attributes. The next level of analysis involved searching, reviewing and naming of themes. The final phase involved a second round of verification and the creation of an implication matrix (an asymmetric matrix that illustrates the number of times each concept was mentioned and the number of times each concept led to each of the other concepts in participants' ladders (Klenosky, 2002)). Then, this matrix was used to construct a useful and informative Hierarchical Value Map [HVM]; which is a chart that illustrates the main links among the attributes, consequences/benefits, values concepts resulted from the means-end chain theory and its associated laddering technique (Jewell & Crotts, 2001; Klenosky, 2002). A summary of the methods used in the study and their link with the research objectives previously stated are detailed in Table 2.

Ethical considerations

The researcher understands there are limitations regarding the validity of a non-Indigenous person conducting research on Indigenous studies. The researcher is aware that her world-view differs from the Australian Indigenous people's world-view. The researcher considers herself a complex individual who comes from a country that has also experienced colonialism (Mexico) but who has also experienced the challenge of having to adopt other culture's values and ways of living (when migrating to Australia); while conserving her Mexican identity. The researcher does not intent to speak for the Australian Indigenous people; but to present visitor perspectives which could be useful for current Indigenous tourism operators and communities considering embarking into tourism.

Through this study, some level of Indigenous participation was achieved as the research design was consulted with the Director of Moondani Balluk's Director and funded by Moondani Balluk Indigenous Academic Unit, Victoria University. Although, the participants of the research project were visitors, the AIATSIS (2012) principles were followed as the focus of the project was on Indigenous tourism. As such, Indigenous tourism stakeholders were contacted before the research started and explained the project, requested for their involvement and agreement, and also explained the benefits of the research. Upon completion of the research, a report with findings was prepared and delivered to them.

Results and discussion

This paper focuses on reporting the results from the semi-structured interviews using the ranking-sorting photo-based technique in regard to the links between the perceived attributes of Indigenous cultural heritage portrayed (traditional and/or contemporary) of two Indigenous tourism activities, and visitor intention to participate. Detailed results from analysing the demographics, psychographics and travel behaviour data are not included in this chapter due to limited space. However, an overview of the participants is included below.

Demographic profile of participants

Table 3 summarises the demographic profile of participants in the present study and also provides statistics for the area based on a report conducted by Tourism and Events Queensland (2017). The sample of the present study represents a relatively well-balanced spread of gender, age and type of visitor. However, it appears that the type of visitor and main reason to travel differ slightly in comparison to the Tourism and Events Queensland's (2017) report. In addition, the sample might be

Table 2. Summary of the methods used in the study.

Method	Research objective	Type of data collected	Deployment	Data analysis	Limitations
Photo-based ranking-sorting procedure	(1) To gain information about visitor intention to participate in Indigenous tourism activities. (1) To compare preferences for tourism activities versus other tourism activities available at the destination.	Quantitative	Participants were given ten photographs and were asked to place the photographs by rank-ordering according with their preference to participate in the activities. The researcher then gave participants another set of the same photographs and asked the participants to sort the photographs according with their intention to participate in the tourism activities while travelling at Cairns. They were asked to place the photographs in two columns titled activities I 'intend to do' or 'do not intend to do'. Participants were then asked to rank the activities in the 'intend to do' column according to their preferences.	Mean rankings were computed for the ranking data to obtain preferences for tourism activities (values were assigned from 1 to 10, with 10 being the most preferable activity). Frequency distribution was computed for the intention to participate in the different tourism activities.	There could be limitations regarding the interpretation of the tourism activity represented on the photographs. A small caption of the tourism activity represented was included to diminish this. Time consuming for participants to understand a new methodology.
Semi-structured interviews using photo-elicitation	(2) & (3) To uncover perceived attributes of Indigenous cultural heritage portrayed in two Indigenous tourism activities. (3) To uncover perceived consequences/benefits, values and constraints regarding intention to participate in Indigenous tourism activities.	Qualitative	Participants were asked to explain their reasons for participating, or not, in each tourism activity. The researcher encouraged the participants to list multiple factors. These responses are typically referred to as attributes (or pull factors). Next, the researcher, using the laddering technique, asked each participant: why is this important to you? This was done for each attribute mentioned and was continued until the participant could not provide any further reasons. This process was undertaken to elicit the consequences/benefits and values (or push factors) related to intention/non-intention to participate.	Coding of the attributes was conducted by an inductive approach (from the data). The coding at the consequences/benefits level followed the pool of motivations used in Pearce and Lee's (2005) study. The coding at the value level followed literature on human values (Kahle, 1983; Rokeach, 1973). To code the constraints, the typology of Crawford and Godbey (1987) was used along with the pool of constraints concepts previously identified in tourism studies and an inductive approach.	The level of depth was limited by the time available for interviews because the sample was made up of actual visitors travelling. It is also recognised that the level of depth was limited to the participants' awareness of their own motivations and constraints

Table 3. Overview of participants at Cairns.

Characteristic	Group	Sample present study (percentage) $n = 209$	Tourism & Events Queensland (2017)
Gender	Female	57	
	Male	43	
Age group	15–24	16	
	25–34	22	
	35–44	13	
	45–54	12	
	55–64	17	
	65+	20	
Type of visitor	Domestic	57	67
	International	43	33
State of residency (domestic visitors)	Victoria	27	25
	New South Wales	27	32
	Queensland	24	23
	Other States/Territories	23	20
Country of origin (international visitors)	United Kingdom	18	
	Other Europe	18	
	France	17	
	Germany	16	
	United States & Canada	8	
	Netherlands	7	
	Italy	6	
	Asian countries	6	
	New Zealand	4	
Main reason to travel to Cairns	Holiday	79	63
	Visiting friends or relatives	7	16
	Business	4	23
	Education	3	
	Other	6	

under represented in terms of visitors from China and Japan, which are the main international markets for Tropical North Queensland (Tourism and Events Queensland, 2017). This might be because they have either not accepted to participate in this study due to language barriers or do not visit visitor centres as they may be travelling in a package tour.

Preference and intention to participate in tourism activities

Figures 1 and 2 summarise the data pertinent to the preferences for, and intention to, participating in Indigenous tourism activities respectively. Figure 1 shows the mean ranking results of the preference for tourism activities and Figure 2 displays the percentage of participants that positioned each activity as an 'intend to participate' activity while in Cairns. For example, Figure 1 illustrates that the most preferable activity to do was 'sightseeing' with a mean score of 8.1 from a possible of 10.0. The Indigenous tourism activities ('short tour' and 'performance') were ranked in the 5th and 7th positions respectively in terms of 'preference'; however, they dropped positions in terms of 'intention to participate' (see Figure 2). It can be seen that only 37% of participants said that they would participate in the 'short tour' and 30% in the performance. Ruhanen et al.'s (2015b) findings claim that visitor intentions (or interest) do not always convert into participation. Therefore, it is suggested that the percentage of visitors actually participating in these activities is less than the percentage reported in the findings of the present study.

Indigenous tourism attributes and links with intention to participate

As previously mentioned, during the semi-structured interviews, the researcher was guided by the means-end chain theory (Jiang et al., 2015) to understand visitor perceptions of the Indigenous tourism activities' attributes and their link with visitor intention to participate in the two tourism

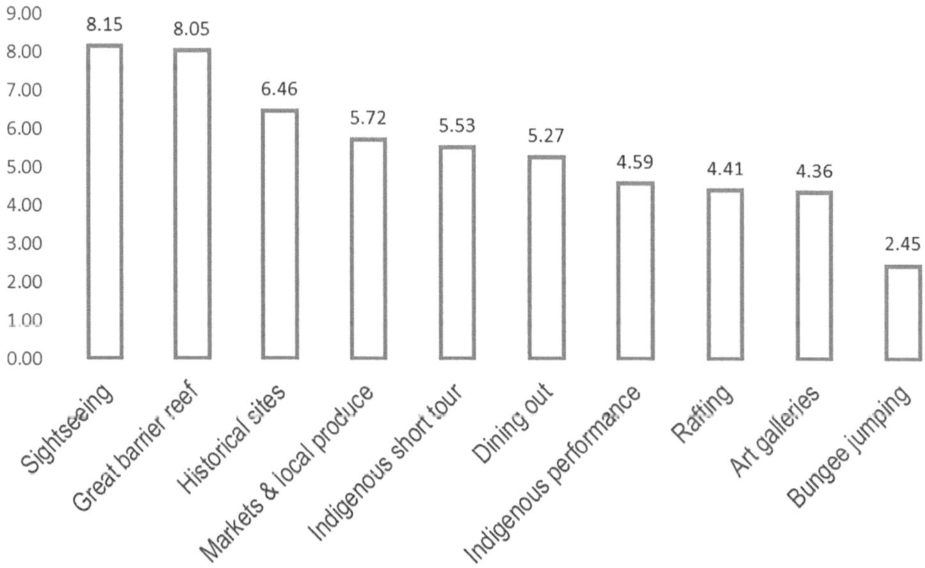

Figure 1. Mean ranking of the preference for tourism activities at Cairns.

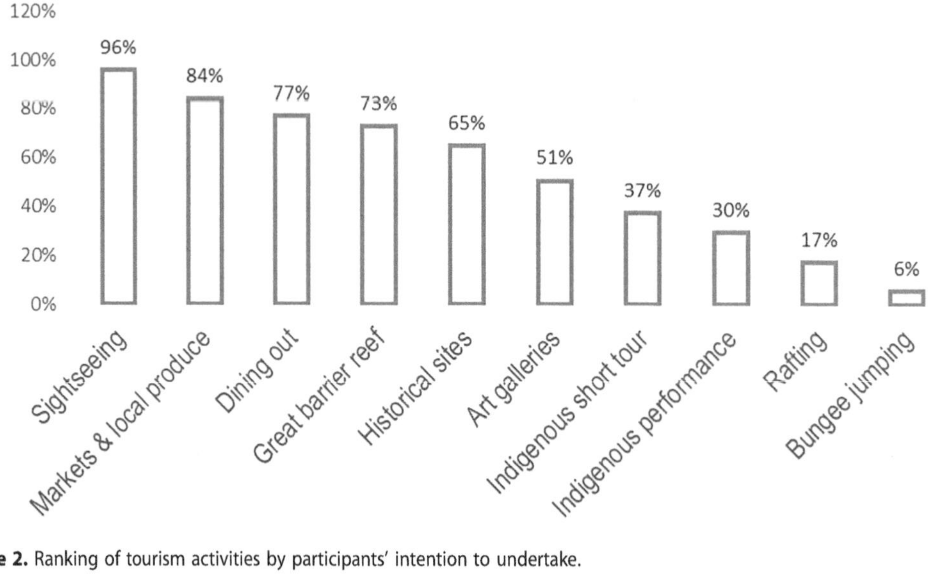

Figure 2. Ranking of tourism activities by participants' intention to undertake.

activities under study. This technique allowed the identification of attributes, consequences/benefits, values and constraints. The results are detailed in the following sections.

Indigenous short tour

The concepts identified on the laddering technique for the Indigenous short tour are summarised in the form of an implication matrix in Table 4. The numbers represented on the matrix are the number of participants that mentioned the concept and the association between attributes, consequences/benefits, and values. For example, the table shows that attribute 5 (A5: History/arts/culture) led to consequence/benefit 2 (C2: Experience differences) 40 times; similarly, consequence 2 was related to consequence 8 (C8: Understanding other culture) 37 times, and so forth.

Table 4. Implication matrix for the short tour.

Attributes	*n*	A1	A2	A3	A4	A5	A6	A7	A8	C1	C2	C3	C4	C5	C6	C7	C8	C9	C10	V1	V2	V3	V4	V5	V6
A1: Authenticity	3			1		3	1		1		1	1			1		2					1	2	1	
A2: Convenient	4					3					1					2	1								
A3: Good quality	4					1	1		1		1						1								
A4: Interaction/hands on	11					8		1			7	1		4			9	1					11	1	3
A5: History/art/culture	65						11	5	6		40	6	6	19	4	10	44	1	2		11	4	45	13	7
A6: Local	13							1	2		5	1	1	5	1		7				1		7	1	
A7: Connection with nature/outdoors	7									4	2	1		1			5			1			3	3	
A8: Unique/famous	9										3	1					4						4		
Consequences/benefits																									
C1: Being close to nature	8										5	1		3			7			1			5	3	
C2: Experience differences	47											4	4	16		3	37	2			6		36	9	5
C3: Gaining a new perspective on life	6												1	1			6					1		6	1
C4: Gaining appreciation	7													3	1	1	4				2		5	4	1
C5: Developing my knowledge	24														2	1	10	2					23		
C6: Support	5																1					4			2
C7: Understanding about country/heritage	10																3				7		8		
C8: Understanding of other culture	46																						36	10	7
C9: Developing my skills and abilities	2																						2		
C10: Exploring the unknown	2																								
Values																									
V1: A world of beauty	1																							1	
V2: Belonging	11																						9	1	
V3: Benevolence	4																								2
V4: Self development	51																							8	5
V5: Self fulfilment	14																								2
V6: Warm relationship with others/developing others	8																								

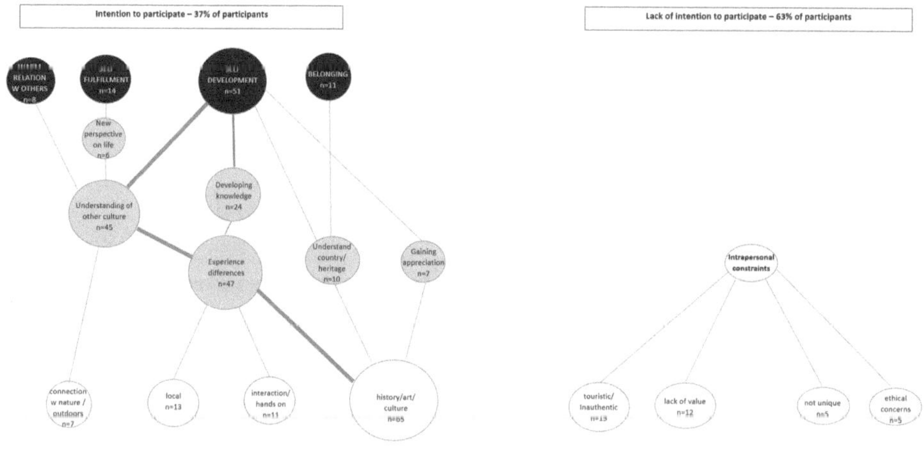

Note: "n" represents the number of participants mentioning the concept. The thickness of the lines reflects the associations between concepts. A thicker line represents a stronger association between concepts.

Figure 3. Hierarchical value map (HVM) for the short tour.

The attributes that were most identified with the short tour are: history/arts/culture, local, and interaction/hands on. Regarding the consequences/benefits, the most identified concepts are: experience differences, understanding other culture, developing knowledge and understanding about country/heritage. Finally, the most identified values are: self-development, self-fulfilment and belonging. The matrix represented in Table 4 was then used to construct a HVM. To decide the cut-off level of the concepts to be included in the HVM for the short tour, a cut-off level of 5 was selected. At this level, 36% of the nonzero cells and 79% of the association between concepts were included. Figure 3 illustrates the final HVM (intention to participate) and the intrapersonal constraints (lack of intention to participate) for the short tour.

The main attribute identified with the short tour was 'history/art/culture'. This is not surprising as Indigenous tourism is based on cultural heritage (Australian Government, 2015). It appears that the majority of the participants ($n = 47$; 64% domestic and 36% international) identified the short tour with a contemporary aspect of the culture (current way of living). However, some participants ($n = 15$; 60% domestic and 40% international) still associated it with a traditional perspective (history).

I am interested in hearing and seeing how Aboriginal people live and do. (Participant 35)

Learn new things. Historical things. Learn the history of the place I am visiting. (Participant 75).

It is evident from Figure 3 that the most dominant set of meanings involved 'history/art/culture' and their link with 'experience differences,' led to the consequence/benefit of 'understanding other culture.' This consequence/benefit was mainly linked to the value 'self-development.' This (learning) has been previously identified as an attractive feature of Indigenous tourism (Abascal et al., 2015; Chang et al., 2006; Kutzner et al., 2009; McIntosh & Johnson, 2005; Notzke, 2004; Ryan & Huyton, 2000; Zeppel, 2002)

Learning about culture and history is interesting <history/art/culture>. *I don't know much about them* <Indigenous people> *except for what I have heard so far* <experience differences>. *They are a big part of Australia, so I think it is fascinating to understand how they lived* <understanding other culture>. *It is always good to learn more about it* <self-development>. (Participant 22)

Other consequences/benefits linked to 'history/art/culture' were 'understanding about country/heritage' and 'gaining appreciation.' The values linked to 'understanding about country/heritage' were 'belonging' and 'self-development.' Participants frequently mentioned the link between Indigenous people and Australia; meaning that the value of 'belonging' could be highly associated with understanding and acknowledging Aboriginal cultural heritage.

> *We like to try to understand their background, their history and culture* <understanding about country/heritage>. *What motivates them? We find them interesting. We should learn about it because it is their country initially* <belonging>. (Participant 118).

'Local' and 'interaction' were another two important identified attributes. These attributes are linked with the consequence/benefit of 'experience differences.' 'Local' is an important attribute as it suggests recognition of diversity. This has been acknowledged as an essential attribute for the long-term sustainability of the sector and as a way to avoid stereotypes (Abascal et al., 2015; Amoamo & Thompson, 2010; Seiver & Matthews, 2016).

> *I haven't had contact with them yet* <interaction/hands on>. *It is just in Australia* <local> ... *So it would be a great opportunity to know about it* <experience differences>. (Participant 146)

The attribute of 'interaction' was previously identified as a positive experience of engaging in Indigenous tourism (Ingram, 2005; Kutzner et al., 2009; McIntosh, 2004; Zeppel, 2002). These previous studies claimed that the level of interaction sought by visitors is not standard among them. In this present study, 'interaction' was mentioned as an opportunity to get in contact with Indigenous people.

> *I haven't had contact with them yet* <interaction/hands on>. *It is just in Australia* <local> ... *So it would be a great opportunity to know about it* <experience differences>. (Participant 146)

'Connection with nature/outdoors' has been identified previously as an important feature for Indigenous tourism attractiveness, and it has been claimed that visitors are more interested in this feature than the culture per se (|Ryan & Huyton, 2000, 2002). Indeed, Abascal et al. (2016) suggested that for some visitors, their lack of intention to participate in Indigenous tourism is because they do not perceive any link between cultural activities and nature-based environments. In the present study, this attribute had a different connotation as participants perceived Indigenous culture as being connected with nature.

> *To see the nature* <connection with nature/outdoors > *and what it actually means to the Aboriginal culture* <understanding other culture>. *To get an idea of the culture of the local area from the Aboriginal point of view* <self-development>. (Participant 30)

Regarding constraints to participate in the short tour, perceived 'authenticity' appears to be the most important intrapersonal constraint. This has been previously identified by Ramkissoon and Uysal (2011). It appears that, for visitors, the fact that an Indigenous activity is developed for tourism purposes makes it 'fake'. It suggests that these visitors are not considering that every person and culture are complex individuals that are defined by their relations: 'a relation will never combine participating bodies in their entirety, but only partially and contextually' (Bignall, 2014, p. 350).

> *Just for tourists. Not real. It is our 4th time here and we have travelled in Aboriginal communities and they are completely different than the tours.* (Participant 91)

'Lack of value' was another significant intrapersonal constraint. Reasons for this include the perceptions that the activity does not add value to the current knowledge of the participant and/or participants can get the knowledge themselves.

> *We come from a town that has relatively high Aboriginal content. We have seen Aboriginal things in Cape York and actually had Aboriginals who explained things to us and things like that, so I think I don't need to go to any Aboriginal tour.* (Participant 4)

> *Rather go to a museum or gallery to see and understand ourselves.* (Participant 187)

Indigenous performance

Table 5 summarises the concepts identified on the laddering technique for the Indigenous performance at Cairns. The attributes that were most identified with this activity are: history/arts/culture,

Table 5. Implication matrix for the performance.

Attributes	*n*	A1	A2	A3	A4	A5	A6	A7	A8	C1	C2	C3	C4	C5	C6	C7	C8	C9	C10	V1	V2	V3	V4	V5	V6	V7
A1: Authenticity	2			1	1	2					1						1		1			1				
A2: Convenient	7					4					1								2							
A3: Good quality	11				1	10			1		4	3	1						2	1			3			
A4: Interaction/hands on	2					2					2								2					1	1	1
A5: History/art/culture	51						4	1	5		35			1	5	10	3	8	30	1	5	4	4	30	9	9
A6: Local	9								3		5				1	2	1	1	7					6	1	
A7: Connection with nature/outdoors	1									1									1					1	1	
A8: Unique/famous	6										5				1				2				1	2		
Consequences/benefits																										
C1: Being close to nature	1										1								1						1	
C2: Experience differences	36											1			3	4		4	29		2	1	2	22	7	5
C3: Have fun	3												1						1				3			
C4: Feeling the special atmosphere	1																						1			
C5: Gaining a new perspective on life	1																		1						1	
C6: Gaining appreciation	6															2	1		2		1		1	3	2	3
C7: Developing my knowledge	14																2	3	2	1				12	1	
C8: Support	4																		1			3				1
C9: Understanding about country/heritage	9																		4		4			8	2	
C10: Understanding other culture	37																				2	1	1	23	6	7
Values																										
V1: A world of beauty	1																									
V2: Belonging	6																							3		1
V3: Benevolence	4																									2
V4: Fun and enjoyment	4																									
V5: Self development	34																								5	1
V6: Self fulfilment	9																									2
V7: Warm relationship with others/developing others	10																									

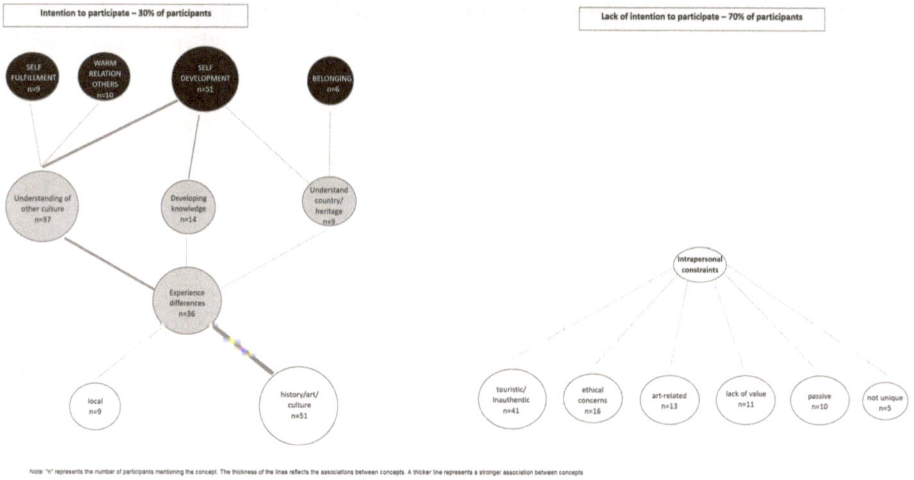

Figure 4. Hierarchical value map (HVM) for the performance.

good quality and local. Regarding the consequences/benefits, the most identified concepts are: understanding other culture, experience differences and developing knowledge. Finally, the most identified values are: self-development and warm relationship with others/developing others. To decide the cut-off level of the concepts to be included in the HVM, a cut-off level of 4 was selected. At this level, 30% of the nonzero cells and 74% of the association between concepts were included.

Figure 4 illustrates the final HVM (intention to participate) and the intrapersonal constraints (lack of intention to participate) for the performance.

Similar to the short tour, the main attribute identified with the performance was 'history/art/culture'. It appears that the majority of the participants ($n = 37$; 68% domestic and 32% international) identified the performance with a contemporary aspect of the culture. However, still some participants ($n = 10$; 80% domestic and 20% international) mentioned a traditional perspective. This contradicts previous findings by Ashwell (2014), who found high levels of awareness of art and craft and its association with historical and stereotypical themes. See quotes below:

> *I think it is very interesting and important to appreciate our Indigenous culture, our Indigenous people.* (Participant 61)

> *I am still finding it very fascinating the way they used to live in the past. The performances are very good, very unique.* (Participant 42)

'Local' was the other attribute that was mentioned for the two Indigenous activities. This concept was identified not only by international visitors (which may be expected as Indigenous people are local to Australia), but also by domestic visitors who seem to be able to distinguish between different Indigenous groups around Australia. This has been previously identified by Galliford (2009) and Ruhanen et al. (2015b) who suggested that domestic visitors have a more realistic understanding of Indigenous people as a current cultural group and are more aware of the Indigenous tourism activities available in Australia.

> *For the culture and history of the country and the people <history/art/culture>. For the traditions of the tribes <local>.*

It is evident from Figure 4 that the most dominant set of meanings involved 'history/art/culture' and 'local' and their link with 'experience differences', which then led to the consequence/benefit of 'understanding other culture'. Participants that mentioned 'understanding other culture' as a consequence/benefit for participating in Indigenous tourism were interested in understanding either the current situation of Indigenous people, or the way they used to live in the past. This consequence/benefit was mainly linked to the value of 'self-development'. However, other values linked to this consequence/benefit are 'self-fulfilment' and 'warm relationship with others/developing others.'

> *I am very interested in lots of different cultures* <history/art/culture>. *I will take any opportunity to learn about different cultures* <understanding other culture>. *To reflect on what I believe* <gaining new perspective on life > *and make me more open minded* <self-fulfilment>. (Participant 43)

> *Very interested in learning their culture and ways* <history/art/culture> <understanding other culture>. *Respecting their culture* <warm relationship with others/developing others>. (Participant 94)

In addition, 'experience differences' was also linked to the consequence/benefit of 'understanding country/heritage'. This consequence/benefit was linked to the value of 'belonging'. It appears that there were two connotations regarding this value. One was the feeling of belonging to Australia, mainly from domestic participants. They mentioned their willingness to understand/learn about Indigenous culture, as part of learning about their own country. The other meaning of this concept was the recognition of Indigenous people as being part of Australia. Previous studies have suggested that participating in Indigenous tourism could strengthen national identity and reconciliation (Galliford, 2009; Higgins-Desbiolles, 2003).

> *Interested in Aboriginal culture* <experience differences>. *It is really the history of Australia and the whole development of Australia it has been around about Aboriginal people* <understanding country/heritage>. *It is where Australia come from* <belonging>. (Participant 41)

As previously mentioned for the short tour, perceived authenticity appears to be the most important intrapersonal constraint to participate in Indigenous tourism. However, the number of participants mentioning this constraint was larger for the performance than for the short tour. Another very important constraint to participate in the performance was 'discomfort/ethical concerns'. It appears that the idea of an Indigenous person performing a show designed specifically for tourism makes some visitors feel a level of discomfort and to express ethical concerns about it. Both of these constraints could be linked to concepts such as commodification of culture, power relations, and the controversy of authenticity in a capital system (Amoamo & Thompson, 2010; Koot, 2016; Ryan & Aicken, 2005; Whitford & Ruhanen, 2016).

> *There is not enough information, so for example if they explain why they are doing it? What does it represent? Because it seems it is just for the tourism sake. If it comes from them, and why they do it, it is good.* (Participant 40)

> *I am not a fan of cultural appropriation for your own benefit. Exploit someone else for culture for your own benefit.* (Participant 136)

> *I don't like them to make their ritual and traditional stuff a commodity.* (Participant 206)

Conclusions and implications for Indigenous cultural heritage management in tourism

This study investigated the link between visitor perceptions of the cultural heritage attributes of two Indigenous tourism activities and their intention to participate in them while travelling in Cairns. In general, the findings suggest that Indigenous tourism could foster cross-cultural understanding; however, there are several constraints to participate. The findings propose that some elements are particularly relevant for the local Indigenous tourism sector as they are shared between the activities. For example, identified elements at the (1) attribute level include: history/art/culture and local; (2) consequence/benefit level include: experience differences, understanding other culture, developing knowledge and understanding about country/heritage; and (3) values level include: self-development and belonging. Constraints include: lack of authenticity, discomfort/ethical concerns, lack of value and not unique. However, not all the concepts were identified for the two activities under investigation. This implies that subsequent research and managerial decisions should be activity specific, rather than considering Indigenous tourism from a general perspective. These findings are important when developing appropriate strategies designed not only to increase the economic and sociocultural benefits of tourism but also to manage cultural heritage. For example, when developing and promoting tourism activities, it is important to celebrate cultural differences and highlight the high-quality of the activity (e.g. local, small groups, specialties, unique, interaction with visitors, etc.). Hence, the identification and inclusion of points of differentiation between Indigenous groups could be done by highlighting the key elements of the local Indigenous culture. This could increase the appeal and sense of 'experience differences,' 'authenticity,' and 'appreciation.' However, when recognising the diversity of Indigenous cultures, it is important to highlight the complexity of the cultures and avoid stereotyped images.

The findings also suggest that the perception of Indigenous culture in Australia is mixed. While some visitors associated it with a traditional past, a large proportion expressed a contemporary perception of Aboriginal culture. Therefore, it is important that Indigenous tourism development and promotion strategies emphasise that Aboriginal culture, as a living culture, is continuously evolving. This could help to increase the visitors' recognition of the Indigenous people and their culture to foster appreciation and respect.

Finally, it also appears that cultural heritage (in particular intangible heritage) could be in danger of being susceptible to commodification and loss of authenticity in order to accommodate visitor expectations. It appears that when visitors are interested in Indigenous tourism, but the perception of authenticity is a constraint, it could either be because visitors are looking for Indigenous people and communities engaging in a complex relation, or because of the stereotypical awareness regarding Indigenous tourism. Hence, creating strategies to manage the tourism interaction as a function of context in conjunction with the Indigenous tourism providers and communities could help with the management of visitor expectations. It is also important to avoid using cliché images and to increase awareness regarding the control by, participation of, and/or benefits for, Indigenous people in the offering of Indigenous tourism. This is needed to avoid beliefs regarding commodification of culture as it appears to be a concern for some visitors.

Acknowledgements

The author would like to acknowledge the Moondani Balluk Indigenous Academic Unit, Victoria University for providing financial support for the fieldwork. The author also gratefully acknowledges the visitor information centre at Cairns, QLD for allowing me to conduct the data collection within its premises.

Disclosure statement

No potential conflict of interest was reported by the author.

Funding

This work was supported by the Victoria University Indigenous Central Research Grant.

References

Abascal, T. E., Fluker, M., & Jiang, M. (2015). Domestic demand for Indigenous tourism in Australia: Understanding motivations, barriers, and implications for future development. *Journal of Heritage Tourism*, *10*(1), 1–20.

Abascal, T. E., Fluker, M., & Jiang, M. (2016). Domestic demand for Indigenous tourism in Australia: Understanding intention to participate. *Journal of Sustainable Tourism*, *24*(8–9), 1350–1368.

Abascal, T. E., Fluker, M., & Jiang, M. (2018). Image-use in tourism research methodology. In W. Hillman & K. Radel (Eds.), *Qualitative research methods in tourism – a handbook* (pp. 248–278). Channel View Publications.

Amoamo, M., & Thompson, A. (2010). (Re)imaging Maori tourism: Representation and cultural hybridity in postcolonial New Zealand. *Tourist Studies*, *10*(1), 35–55. doi:10.1177/1468797610390989

Ashwell, J. (2014). *Perceptions and attitudes of international tourists towards Aboriginal tourism*. (Unpublished doctoral dissertation), Flinders University, Adelaide, Australia.

Australian Government. (2015). *Australian Indigenous cultural heritage*. Retrieved from http://www.australia.gov.au/about-australia/australian-story/austn-indigenous-cultural-heritage

Australian Institute of Aboriginal and Torres Strait Islander Studies. (2012). *Guidelines for ethical research in Australia Indigenous studies*. Canberra. Retrieved from http://aiatsis.gov.au/research/ethical-research/guidelines-ethical-research-australian-indigenous-studies

Bignall, S. (2014). The collaborative struggle for excolonialism. *Settler Colonial Studies*, *4*(4), 340–356. doi:10.1080/2201473x.2014.911651

Butler, R., & Hinch, T. (2007). *Tourism and indigenous peoples: Issues and implications* (1st ed.). Amsterdam/Boston: Elsevier/Butterworth-Heinemann.

Butler, S. (2008). Mall intercept survey. In D. P. J. J. Lavrakas (Ed.), *Encyclopedia of survey research methods* (pp. 448–449). Thousand Oaks, CA: Sage.

Chan, J. H., Iankova, K., Zhang, Y., McDonald, T., & Qi, X. (2016). The role of self-gentrification in sustainable tourism: Indigenous entrepreneurship at Honghe Hani rice terraces world heritage site, China. *Journal of Sustainable Tourism*, *24*(8–9), 1262–1279. doi:10.1080/09669582.2016.1189923

Chang, J., Wall, G., & Chu, S.-T. (2006). Novelty seeking at aboriginal attractions. *Annals of Tourism Research*, *33*(3), 729–747. doi:10.1016/j.annals.2006.03.013

Coshall, J. T. (2000). Measurement of tourists' images: The repertory grid approach. *Journal of Travel Research*, *39*(1), 85–89. doi:10.1177/004728750003900111

Crawford, D. W., & Godbey, G. (1987). Reconceptualizing barriers to family leisure. *Leisure Sciences*, *9*(2), 119–127. doi:10.1080/01490408709512151

Creswell, J. W., & Plano Clark, V. L. (2011). *Designing and conducting mixed methods research* (2nd ed.). Thousand Oaks, CA: Sage.

Dyer, P., Aberdeen, L., & Schuler, S. (2003). Tourism impacts on an Australian indigenous community: A djabugay case study. *Tourism Management*, *24*(1), 83–95. doi:10.1016/S0261-5177(02)00049-3

Galliford, M. (2009). *Transforming the tourist: Aboriginal tourism as investment in cultural transversality*. (Unpublished doctoral dissertation), University of South Australia, Adelaide, Australia.

Gutman, J. (1982). A means-end chain model based on consumer categorization processes. *Journal of Marketing*, *46* (2), 60–72.

Haas, G. E., Driver, B. L., & Brown, P. J. (1980). *Measuring wilderness recreation experiences*. The Wilderness Psychology Group Annual Conference (pp. 20–40). Durham: University of New Hampshire.

Higgins-Desbiolles, F. (2003). Reconciliation tourism: Tourism healing divided societies. *Tourism Recreation Research*, *28*(3), 35–44.

Hodgson, R., Firth, T., & Presbury, R. (2007). Aboriginal cultural tourism in the northern territory: Sociocultural impacts of tourism on the Manyallaluk community. In J. Buultjens & D. Fuller (Eds.), *Striving for sustainability: Case studies in Indigenous tourism* (pp. 291–332). Lismore: Southern Cross University Press.

Ingram, G. (2005). A phenomenological investigation of tourists' experience of Australian indigenous culture. In C. Ryan & M. Aicken (Eds.), *Indigenous tourism: The commodification and management of culture* (1st ed., pp. 21–34). Amsterdam: Elsevier.

Jewell, B., & Crotts, J. C. (2001). Adding psychological value to heritage tourism experiences. *Journal of Travel & Tourism Marketing*, *11*(4), 13–28. doi:10.1300/J073v11n04_02

Jiang, S., Scott, N., & Ding, P. (2015). Using means-end chain theory to explore travel motivation: An examination of Chinese outbound tourists. *Journal of Vacation Marketing*, *21*(1), 87–100. doi:10.1177/1356766714535599

Jones Donald Strategic Partners. (2009). *Project Indigenous tourism: A research based consultancy to assist TA/RET to provide marketing guidance to promoters and suppliers of Indigenous tourism experiences for the domestic market*. Unpublished manuscript.

Kahle, L. R. (1983). *Social values and social change: Adaptation to life in America*. New York, NY: Praeger.

Klenosky, D. B. (2002). The 'pull' of tourism destinations: A means-end investigation. *Journal of Travel Research*, *40*, 385–395.

Koot, S. P. (2016). Contradictions of capitalism in the South African Kalahari: Indigenous Bushmen, their brand and Baasskap in tourism. *Journal of Sustainable Tourism*, *24*(8–9), 1211–1226. doi:10.1080/09669582.2016.1158825

Kutzner, D., Wright, P. A., & Stark, A. (2009). Identifying tourists' preferences for aboriginal tourism product features: Implications for a northern first nation in British Columbia. *Journal of Ecotourism*, *8*(2), 99–114. doi:10.1080/14724040802695991

McIntosh, A. J. (2004). Tourists' appreciation of Maori culture in New Zealand. *Tourism Management*, *25*(1), 1–15. doi:10.1016/s0261-5177(03)00058-x

McIntosh, A. J., & Johnson, H. (2005). Understanding the nature of the Marae experience: Views from hosts and visitors at the Nga Hau E Wha national Marae, Christchurch, New Zealand. In C. Ryan & M. Aicken (Eds.), *Indigenous tourism: The commodification and management of culture* (pp. 35–50). Amsterdam: Elsevier.

Mkono, M. (2016). Sustainability and Indigenous tourism insights from social media: Worldview differences, cultural friction and negotiation. *Journal of Sustainable Tourism*, *24*(8–9), 1315–1330. doi:10.1080/09669582.2016.1177066

Muller, D. K., & Pettersson, R. (2005). What and where is the indigenous at an indigenous festival? Observations from the winter festival in Jokkmokk, Sweden. In C. Ryan & M. Aicken (Eds.), *Indigenous tourism: The commodification and management of culture* (1st ed., pp. 21–34). Amsterdam: Elsevier.

Notzke, C. (2004). Indigenous tourism development in Southern Alberta, Canada: Tentative Engagement. *Journal of Sustainable Tourism*, *12*(1), 29–54.

Pearce, P. L., & Lee, U.-I. (2005). Developing the travel career approach to tourist motivation. *Journal of Travel Research*, *43*(3), 226–237. doi:10.1177/0047287504272020

Pomering, A., & White, L. (2011). The portrayal of indigenous identity in Australian tourism brand advertising: Engendering an image of extraordinary reality or staged authenticity? *Place Branding and Public Diplomacy*, *7* (3), 165–174. doi:10.1057/pb.2011.19

Ramkissoon, H., & Uysal, M. (2011). The effects of perceived authenticity, information search behaviour, motivation and destination imagery on cultural behavioural intentions of tourists. *Current Issues in Tourism*, *14*(6), 537–562.

Rokeach, M. (1973). *The nature of human values*. New York, NY: Free Press.

Ruhanen, L., Whitford, M., & McLennan, C.-L. (2015a). Exploring Chinese visitor demand for Australia's Indigenous tourism experiences. *Journal of Hospitality and Tourism Management*, *24*, 25–34. doi:10.1016/j.jhtm.2015.07.002

Ruhanen, L., Whitford, M., & McLennan, C.-L. (2015b). Indigenous tourism in Australia: Time for a reality check. *Tourism Management*, *48*, 73–83. doi:10.1016/j.tourman.2014.10.017

Ryan, C. (2005). Who manages indigenous cultural tourism product – aspiration and legitimization. In C. Ryan & M. Aicken (Eds.), *Indigenous tourism: The commodification and management of culture* (1st ed., pp. 21–34). Amsterdam: Elsevier.

Ryan, C., & Aicken, M. (2005). *Indigenous tourism. The commodification and management of culture* (1st ed.). Amsterdam: Elsevier.

Ryan, C., & Huyton, J. (2000). Who is interested in aboriginal tourism in the northern territory, Australia*? A cluster analysis. Journal of Sustainable Tourism*, *8*(1), 53–88. doi:10.1080/09669580008667349

Ryan, C., & Huyton, J. (2002). Tourists and aboriginal people. *Annals of Tourism Research*, *29*(3), 631–647.

Seiver, B., & Matthews, A. (2016). Beyond whiteness: A comparative analysis of representations of aboriginality in tourism destination images in New South Wales, Australia. *Journal of Sustainable Tourism*, *24*(8–9), 1298–1314. doi:10.1080/09669582.2016.1182537

Tian, S., Crompton, J. L., & Witt, P. A. (1996). Integrating constraints and benefits to identify responsive target markets for museum attractions. *Journal of Travel Research*, *35*(2), 34–45.

Tourism and Events Queensland. (2017). *Tropical North Queensland regional snapshot: Year ending March 2017*. Retrieved from https://cdn1-teq.queensland.com/~/media/5fe9b37cfd304c30a0835b70214c041a.ashx?vs=1&d=20170803T182942

Tourism Research Australia. (2016). *National and international visitor survey*. Retrieved from Tourism Research Australia Online: http://traonline.tra.gov.au.

Walker, K., & Moscardo, G. (2016). Moving beyond sense of place to care of place: The role of indigenous values and interpretation in promoting transformative change in tourists' place images and personal values. *Journal of Sustainable Tourism*, *24*(8–9), 1243–1261. doi:10.1080/09669582.2016.1177064

Whitford, M., & Ruhanen, L. (2014). Indigenous tourism businesses: An exploratory study of business owners' perceptions of drivers and inhibitors. *Tourism Recreation Research*, *39*(2), 149–168.

Whitford, M., & Ruhanen, L. (2016). Indigenous tourism research, past and present: Where to from here? *Journal of Sustainable Tourism*, *24*(8–9), 1080–1099. doi:10.1080/09669582.2016.1189925

Zeppel, H. (2002). Cultural tourism at the Cowichan native village, British Columbia. *Journal of Travel Research*, *41*(1), 92–100.

The interpreter as researcher: ethical heritage interpretation in Indigenous contexts

Chance Finegan

ABSTRACT

This paper integrates Indigenous research and heritage interpretation scholarship by addressing how interpretation should be underpinned by research congruent with Indigenous expectations for research ethics and methods. Program research and delivery is a profound exercise of power, for interpreters mold public opinion. In a settler-colonial or Indigenous context, critically analyzing sources for one's interpretative programs and media is particularly imperative, as Western researchers have historically been on colonialism's front lines. The interpretation of Indigenous themes is fraught with difficulties, but existing interpretation literature myopically focuses on program delivery and evaluation methods. More attention must be paid by interpreters to the problems of relying on Western, positivist research in Indigenous contexts. This paper considers Indigenous research methods and ethics, provides an overview of the connection between Western knowledge and colonialism, and demonstrates the problems this poses for interpreters working with Indigenous content. I offer concrete suggestions for staff to improve their engagement with Indigenous peoples. Ultimately, I argue, interpreters must become more critical of themselves and their sources to successfully address Indigeneity.

Introduction

This paper will attempt to integrate Indigenous research literature and protected area[1] heritage interpretation practice. I argue that heritage interpreters in protected areas must broaden their self-perceptions to include more than working as simple communicators when they work in settler-colonial settings such as parks in Canada and the United States (nation-states governed by colonists who intended to stay and replace Indigenous peoples; see Veracini (2010), Wolfe (2006), or Tuck and Yang (2012) for an overview of settler-colonialism). Interpreters are also researchers, stewards of stories, and people with the power to give credence to certain knowledge. In an Indigenous or settler-colonial context, these roles are of paramount importance and carry additional weight, as interpreters must navigate the boundary between the colonized and the colonizer. Building a culture of criticality (i.e. interrogating oneself and one's sources) may help to resolve the difficulties associated with interpretation at this intersection.

Heritage interpreters forge connections between visitors and natural and cultural heritage resources at places like parks and protected areas, historic sites, nature centers, zoos, and aquariums. This paper focuses specifically on heritage interpreters employed by settler-colonial park agencies (e.g. the U.S. National Park Service or Parks Canada). I begin by offering a definition of heritage interpretation and a discussion of the interpretation/Indigenous intersection. The following two sections provide an overview of interpretation training and a discussion of the differences between Western and Indigenous knowledge systems. I then offer specific guidelines for interpretation supervisors working in an Indigenous context. The objectives of this article are to (1) focus attention on the role of research in protected area heritage interpretation, (2) create space for thinking about interpretation as an exercise of power that uses information to mold public perceptions, and (3) highlight the intersection between the politics of knowledge creation, settler-colonialism, and protected area heritage interpretation in Indigenous contexts.

Heritage interpreters in protected areas 'directly support the preservation mission' of parks and 'are charged with helping audiences care about park resources ... ' (Bacher et al., 2007, p. 1). Interpretation is, at its core, an attempt to communicate the value of a given resource to an audience (Ham, 1992). Guided walks, formal talks, wayside exhibits, and brochures for self-guided activities are all forms of heritage interpretation.

'The raw material of interpretation is information,' notes Tilden (2009, p. 48), and interpreters depend on 'ammunition' from the academic (research) community to 'tell the visitor what he wants to know' (Tilden, 2009, p. 191). Beck and Cable (2002, p. 20) state interpretation is more than information, but is also focused on 'revealing deeper meaning and truth.' That interpretation centers on revealing meanings (Foxlee, 2007; Moscardo, 2007; Staiff, Bushell, & Kennedy, 2002) is a critical point, for emphasizing the *act* of revealing meaning looks past the *decision* about which meaning to reveal in the first place. Given that landscapes and resources have multiple, culturally specific meanings associated with them (Ashmore, 2004; Clark & Scheiber, 2008; David & Thomas, 2016; Gilmore, 2008), and that interpreters often work in places where these multiple meanings may intersect (such as a site that commemorates a treaty signing), beginning with revealing meaning, rather than contesting and constructing it, is problematic.

This is a crucial difference – if aspiring protected area interpreters are taught they reveal meanings, the implication is that someone else has already decided which values to attach to the resource. This positions interpreters as passive conveyors of information, which they are not. Interpreters are instead researchers, sifting through information to determine what to present to visitors, as well as meaning-makers, who speak with the authority of their agencies and either reinforce or challenge existing perceptions about the resource.

Consider the classic interpreter's creed: 'through interpretation, understanding; through understanding, appreciation; through appreciation, protection' (Tilden, 2009, p. 65). While this is useful, it begins in the middle, ignoring that interpretation begins with knowledge and research. The creed should perhaps begin with something akin to, 'informed by ethical research and knowledge-gathering, multiple meanings are acknowledged and interpreted, through interpretation ... .' Failing to acknowledge that interpreters are meaning-makers is troubling. Interpreters should be fully aware of their power to shape public opinion and knowledge.

Moreover, ignoring the role of research in interpretation suggests that interpreters ought not think critically about how knowledge is constructed and used. Interpreters have a duty to question their sources. This is the crux of the problem this paper interrogates, for interpretation founded on Western academic knowledge and offered within an Indigenous context presents challenges interpreters need to be aware of. Basing interpretation entirely on Western science and knowledge '... communicates an assumption that this knowledge is universal, authoritative, and seemingly, outside the realm of contestation' (Staiff et al., 2002, p. 102). This, of course, is patently untrue as many Indigenous scholars (Disko & Tugendhat, 2014; Kovach, 2010; LaDuke, 2015; Lambert, 2014; McGregor, 2004; Nadasdy, 1999; Smith, 1999) have reminded their Western peers. While the Western settler-colonial project may have succeeded in creating and sustaining settler-colonial governments across the world (e.g. Canada and the USA), multiple epistemologies (i.e. Western and Indigenous) still exist within those territories.

Research and academia have long accompanied imperialism and colonization (Geniusz, 2009; Griffiths & Robin, 1997; Grove, 1996; Kovach, 2010; Lambert, 2014; Smith, 1999). Smith discusses this at length. She notes, 'from the vantage point of the colonized ... the term "research" is inextricably linked to European imperialism and colonialism' (1999, p. 1). Interpreters un/knowingly become complicit in ongoing colonial processes if they are working in an Indigenous context and do not critically question the research on which they base their work and do not represent the multiple meanings that inhabit the resources they interpret.

Heritage interpretation literature and training

Existing interpretation literature focuses strongly on interpretive methods, program demand, and program outcomes (for example, Madin & Fenton, 2004; Ogle, 2016; Powell & Ham, 2008; Ren & Folta, 2016; Tubb, 2003). This suggests the knowledge and epistemology underpinning interpretation are uncontroversial or unworthy of examination. This is unwarranted. How one learns what one interprets is equally as important as the methods for interpreting this knowledge and measuring success in delivering content. In settler-colonial societies, where protected areas preserve state heritage, the intersection of power, knowledge, and race need to be understood by heritage interpreters, as this paper will make clear.

Interpretation is ostensibly an effort to provoke people into thinking about deeper, more complex meanings of resources than they might otherwise consider. But who is assigning those meanings and with what knowledge? To whom are we giving power, intentionally or not, when we choose our sources for programs? It is not my claim that interpreters are strictly limited to using Western knowledge in their work, but rather that heritage interpretation as a profession lacks a culture of criticality about research and sourcing that, together with other factors (e.g. the tendency of museums to reinforce rather than critically examine dominant European narratives (Wall, 2011)), is deeply problematic when working in Indigenous contexts. A failure to emphasize the need to (1) be aware of the problems associated with relying on the academy for knowledge about Indigenous peoples and (2) be critical of one's self and sources is apparent in the training heritage interpreters receive.

The National Association for Interpretation (NAI), based in the United States, offers a variety of professional certifications in heritage interpretation. Its most commonly used program is the certified interpretive guide credential (CIG; NAI/E. Jacobs, personal communication, 14 February 2018). The CIG is 'designed as a training program for those who interpret nature or history with the public ... ' (Brochu & Merriman, 2002, p. 6). The program is particularly important given the U.S. National Park Service's lack of standardized interpretive training for seasonal park ranger-interpreters across its 400-plus protected areas (U.S. National Park Service/K. Olson, personal communication, 12 February 2018). The CIG is an entry-level course designed to help people starting their careers in interpretation (as many seasonal employees are). Additionally, while NAI does not track the employment of its students, over 65% of its general membership comes from the American protected area sector (NAI/E. Jacobs, personal communication, 14 February 2018).

The CIG training textbook (National Association for Interpretation, Brochu & Merriman, 2012) teaches students to identify a resource's tangible and intangible characteristics and to assign those to universal concepts. For example, for an American, a bald eagle might represent universal concepts of freedom and strength. Such instruction is not unreasonable, but the textbook does not expressly discuss that different interpreters will identify different characteristics and concepts.

More importantly, the CIG textbook does not teach trainees to question if they should even be identifying those characteristics and concepts in the first place. As an example, who would I, a white male raised in the southern United States and now living in Toronto, be to earn a job at Pacific Rim National Park Reserve (Vancouver Island, Canada) and attempt to put together a program on European and Nuu-chah-nulth perspectives on salmon? If I chose to do so, what are some of the challenges I would need to navigate? McGregor (2004, p. 394) warns us that Indigenous knowledge '... is not just knowledge *about* the relationships with Creation, it *is* the relationship with Creation; it is the way that one relates.' How could I share with visitors Nuu-chah-nulth knowledge about salmon, therefore? Would it even be possible for me to do so, given that my relationship with salmon is likely very different from a Nuu-chah-nulth person's relationship? The CIG program does not ask students to confront such questions about who should be identifying a resource's universal characteristics.

One might be tempted to answer these questions by suggesting that so long as one is properly informed, offering interpretive programs or media should not present ethical dilemmas. This ignores the larger issue – interpretive training fails to critically examine or teach, with nuance, the process of becoming 'properly informed.' Throughout the CIG program, students learn about the importance of knowing the resource being interpreted. The U.S. National Park Service's interpretive equation, which names (1) knowledge of the resource, (2) knowledge of the audience, and (3) interpretive technique as the three 'ingredients' for interpretation, underscores the importance of knowing about the resource (National Association for Interpretation et al., 2012). But instructing students to simply know the resource without any further comment belies the complicated nature of knowledge.

NAI's CIG instructors tell students to seek out unbiased sources to help develop their programs and media, but the training's discussion of bias is flawed and conflates bias with a conflict of interest. For example, the textbook suggests academic articles are an unbiased source and that trade publications are biased. Consider developing an interpretive program about water pollution. While academics researching pollution may not suffer from the same conflicts of interest as a chemical company employee undertaking similar work, the professors are not free from bias. Their work is informed by their own positionality. It is not that academics are unbiased, it is that they likely suffer from less of a conflict of interest.

In particular, Western approaches to the creation and validation of knowledge are deeply embedded in academia, which is notoriously hostile to Indigenous peoples, their voices, and their epistemologies (Kovach, 2010; Lambert, 2014; Smith, 1999). As Kovach (2010, p. 79) astutely points out, 'much of what dominant society perceives as legitimate knowledge is generated by a rather small, homogeneous group of people in formal institutions … .' Today's research (including this essay) is informed by yesterday's research. Early-career academics are peer-reviewed by veteran researchers. Western knowledge is built brick-by-brick, paper-by-paper, with each one stacked atop the 'accepted' body of knowledge.

'Research,' writes Smith (1999, p. 4), 'is not an innocent, distance exercise, but an activity that has something at stake and that occurs in a set of political and social conditions.' Interpreters rely on the research of others. Without understanding how research under consideration for use in a program may be perceived by Indigenous communities and the role academia has played in the colonization of Indigenous territories (Bewell, 2004; Kovach, 2010; Pratt, 2007; Smith, 1999), interpreters cannot hope to present Indigenous themes in a manner that contributes to settler/Indigenous reconciliation.

When Western researchers, grounded in a written, quantitative, discrete, and positivist epistemology, attempt to consider Indigenous knowledge (generally holistic, oral, and relational), those researchers are translating Indigenous knowledge into a form deemed acceptable by other Western scientists who will peer-review the draft manuscripts. This is a very challenging task, because 'culture holds knowledge, knowledge holds culture; they are iterative, independent, and alive,' (Kovach, 2010, p. 163). In other words, knowledge is culturally embedded. This connection is nearly inescapable. Knowledge derives not just from 'the truth' but from accepted means of discovering the truth and from prior research. Working inside a university generally means working in a context of Western-centric thought. The CIG training, however, does not discuss this and instead relies on a simplistic understanding of bias. The yardstick for evaluating sources appears to be conflicts of interest; the training needs a more fulsome and critical look at 'bias' and the broader systems that underpin knowledge and potential sources. The CIG program should teach students to interrogate all sources critically and to consider how academic knowledge creation is connected to and perpetuates dominant cultures and their power.

Instructions to merely 'gather factual material' (National Association for Interpretation et al., 2012, p. 31) ignore the intersection of knowledge production and settler-colonial power in places such as Canada and the USA. Knowledge creation and distribution is a political act (Kovach, 2010; Smith, 1999). This demands a deeper consideration than the CIG program currently provides.

Research inherently transfers power from study subjects to the researcher (Smith, 1999). Nadasdy (1999, p. 10) helps us to see this by writing that when researchers interview Indigenous peoples, they are 'deciding *for* Elders and hunters which of their experiences are relevant' to the study, because the researchers choose (1) which data to collect and how to do so, (2) how to manipulate and analyze the data, and (3) how to present it to the public. Then, this power is extended and reinforced when interpreters are trained to look towards academics and experts to source their programs and media. Nadasdy gives the illustrative example of maps' authority, which '… is derived not from any inherent "truth" in the maps they [cartographers] produce, but from the fact that navigators have been trained to use and rely on these maps … ' (1999, p. 12). Simply instructing interpreters-in-training to rely on 'the research' without additional conversations about how knowledge is constructed is inappropriate for working in Indigenous contexts.

Some may suggest that interpretation of Indigenous heritage and perspectives is a topic best reserved for advanced trainees. To do so would be myopic in settler-colonial societies such as Australia, Canada, New Zealand, and the United States. All this territory was, prior to colonization, the sovereign domain of Indigenous peoples. In the United States, for example, there is not a single major national park that does not have a close connection to Indigenous peoples (Keller & Turek, 1998). One cannot credibly claim to present the heritage of these places by ignoring Indigenous perspectives and/or the effects of settler-colonialism on Indigenous peoples. I am not suggesting that every park and every interpretive offering necessarily must have an Indigenous focus or component. Instead, I argue that given the fundamental role of settler-colonist/Indigenous interactions in the history and ongoing politics of places like Canada and the USA, interpreters must be equipped from the very start of their careers with a basic understanding of Indigeneity and how it relates to interpretation. It is not enough, nor is it proper, to leave considerations of Indigeneity to advanced training or refresher courses.

Lackey (2008) surveyed 130 universities across North America about their coursework in heritage interpretation. This is the most recent empirical research into interpretive curricula in North America. Lackey reports the most common skill taught in post-secondary interpretation classrooms is the development of programs, talks, and walks. Nowhere in the top 14 responses were issues of research or sourcing for interpretive offerings mentioned.

Lackey also asked faculty members about the textbooks they use to teach interpretive methods. The four most-common books are (in descending order): Ham (1992), Tilden (2009), Beck and Cable (2002), and Knudson, Cable, and Beck (1995). Tilden's *Interpreting Our Heritage* and Ham's *Environmental Interpretation* are complementary. Tilden provides a theoretical introduction to interpretation while Ham provides applied instruction. Yet, these texts are poorly suited to interpretation within Indigenous or settler-colonial contexts. As he first did in the 1950s, Tilden (2009) instructs interpreters to rely on 'the specialists' (p. 48) and, in cases where scientists disagree, to present the various hypotheses so that the visitor may puzzle it out for herself. Further on, Tilden re-iterates the importance of relying on researchers, noting that academics 'reveal evidence' at historic sites, allowing interpreters to 'bring any instances of the historic past so vividly to observers [visitors] that they feel a part of it' (p. 191). But Tilden does not address any of the drawbacks of using historians or archaeologists (e.g. a reliance on Western academia).

For his part, Ham (1992) focuses on the communicative aspect of interpretation, the importance of 'acquiring advanced knowledge in some subject area' (p. xviii), and the ability of interpreters to serve as 'translators … [of] the technical language of a natural science or related field' (p. 3). He refers to the need to 'carefully select facts,' but quickly moves on to discuss interpretive methods without further rumination on how one ought to go about that selection. Perhaps is it unfair to criticize Ham and Tilden, as they are primarily concerned with providing interpreters with the practical advice needed to succeed in their careers. But is a conversation about knowledge systems and being critical of learning from 'specialists' not part of the theory and methods of sound interpretation?

Others may suggest that Ham and Tilden were writing decades ago and are products of their time. While it may be true that Tilden and Ham have been cutting-edge in the past, that does not mean their work should go unquestioned. Put another way, such a defense ignores that Ham and Tilden are products of a Euro-American settler-colonist society, the overriding goal of which is to erase Indigenous peoples and histories (Veracini, 2011) and to become the 'recognized, natural, rightful owners of the land' (Tuck & Gaztambide-Fernandez, 2013, p. 77). As with anyone else, their writing is informed by their positionality (in this case, as settler-colonists). Tilden and Ham can and ought to be critically examined. This is not to say they are without merit, but rather that their shortcomings should be acknowledged and addressed with other materials for interpreters-in-training.

The other texts on Lackey's (2008) most-common North American heritage interpretation textbook list suffer from similar problems. Consider the third-most-used book, Beck and Cable's *Interpretation for the twenty-first Century* (2002). Beck and Cable are nearly silent on issues related to research. While there is a chapter supposedly devoted to considering 'quantity and quality' of facts, it is focused not on how one ought to conduct research for one's program, but instead, on (1) not overwhelming visitors with facts and (2) refuting visitors' misconceptions. This text particularly illuminates gaps in the literature about the park-Indigenous nexus, for Beck and Cable teach that 'the interpreter creates separate lists of tangibles, intangibles, and universal concepts' (p. 24) and interpreters 'shape it [information] into interpretation, revealing meanings associated with the sites … this is a creative and personal process … ' (p. 29). Their emphasis on the deeply personal role of the interpreter is a clear theme throughout the book, with comments such as 'the interpreter creates a whole, yet focused message … ' and 'the interpreter's story becomes that of his or her listeners as well' (p. 38).

This model of interpreters as sculptors of their programs and media must be considered alongside the stark whiteness of protected area professionals. The U.S. National Park Service, for example, is 83% white (Nelson, 2016). If those who drive the messages American (in this case) parks share are overwhelmingly non-Indigenous and if interpretation literature is silent on issues relating to research methods and Western versus Indigenous epistemologies, how can agencies appropriately interpret Indigenous history, heritage, and perspectives? To succeed in presenting the full, complex, nuanced story of the United States, its heritage interpreters need more guidance than to simply listen to 'experts' and to avoid swamping the visitor with extraneous facts.

Contemporary knowledge creation and distribution processes inherently center power in universities. This is the very heart of the matter – interpreters are being superficially taught they are to (1) turn to experts for information and (2) craft a compelling story as a painter does a masterpiece with that information. This works well for a program on, for example, butterfly migration between Canada and Mexico. In an Indigenous context, however, the model breaks down and additional care must be taken. The interpreter must be more critical of the experts they choose to draw upon and ask themselves questions such as: How did the researcher gather this knowledge? How did this research project affirm Indigenous sovereignty rather than harm it? What is my positionality and what internal biases do I hold? Why am I choosing to do this program? Do I have the community's support for offering these teachings? But such questions and the need for criticality are glossed over or ignored in common interpretation textbooks.

The fourth-most-common text on Lackey's list, a 500-page book by Knudson et al. (1995) has a section about Indigenous-focused interpretation. It is a total of two pages plus a paragraph in length (or roughly 0.4% of the book). Given the profound relationship between protected areas and Indigenous peoples (to say nothing of the broader settler/Indigenous relationship and heritage), this is insufficient. Imagine, for example, trying to tell the story of the American South by ignoring Black history and perspectives. As impossible as such a task would be, so too is telling the story of settler-colonial places like Canada or New Zealand without Indigenous perspectives.

Knudson et al. (1995) manage to pack numerous stereotypes into their two pages about Indigenous interpretation, referencing 'the wisdom of the natives [*sic*] who lived in a mostly harmonious relationship with the land' (Knudson et al., 1995, p. 336). This is the ecologically noble savage fallacy, which 'denies the realities of native people's lives, reducing the rich diversity of their beliefs, values, social relations, and practices to a one-dimensional caricature … [and] … is an unattainable ideal'

(Nadasdy, 2005, p. 293). Additionally, the statement is written in the past tense (i.e. 'who lived'). Indigenous peoples still live in North America; using the past tense to describe them suggests either a belief that they are no longer Indigenous (i.e. have fully assimilated – which they have not) or that they no longer exist (simply untrue). Either way, it is problematic – as is that the section on Indigenous interpretation is in a chapter on 'historical interpretation.' To be very clear: Indigenous nations, culture, and heritage are living to this very day, despite the settler-colonists' efforts, for centuries, to stamp out Indigeneity through an ongoing genocide (Wolfe, 2006). Indigenous peoples should not be written about in the past tense, nor should aspiring interpreters be trained to think of Indigenous peoples as on the same conceptual plane as, say, the French-Canadian fur traders or the Confederate States of America (both historical groups of people).

Later in their section on Indigenous interpretation, Knudson et al. (1995) do offer advice for the interpretation of Indigenous themes, heritage, and perspectives in the form of a six-bullet list. While the instructions they give are useful (e.g. 'Involve the people in the planning and presenting of their own culture' (Knudson et al., 1995, p. 337)), six bullets out of a 500+ page textbook are not enough to overcome the flaws throughout the rest of the book. Their comments are weakened further by putting Indigeneity into a literal box on the page. In settler-colonial nations, protected area interpreters cannot and should not view Indigeneity as something that belongs in its own box and space, removed from other considerations. The heritage of settler-colonial nation-states is as much an Indigenous story as a settler story; the two cultures' stories and histories are inseparable. To put Indigenous peoples into their own box and file them away as merely historical interpretation is damaging.

Interpreters need to be familiar with the challenges of working in Indigenous contexts, but rather than directly address this, Knudson, Cable, and Beck simply breeze past it with comments like 'Facing these [controversial] issues requires careful preparation of factual presentations … .' (1995, p. 147). Who determines what the facts are? How does one find the facts? Whose knowledge is valid? Such questions are left unanswered by the authors. The book lacks a discussion about research for programs. 'Research' in the index refers the reader to passages where research *about* interpretation is discussed in the text, not to passages guiding students on research *for* interpretation. This is troubling, for it implies an unquestioned reliance on universities. Interpretation, particularly of settler-colonial or Indigenous heritage, is messy. But Knudson, Cable, and Beck do not address such complexity in their book. They simply take it for granted (if the authors considered it at all) that interpreters-in-training who read their textbook will understand that seeking out multiple perspectives, including those outside of the academy, is necessary to ethically identify and understand the story of a place or resource.

Putting aside questions of research methods instruction, the authors' stereotypes of and outright falsehoods about Indigenous peoples should immediately disqualify the text from being used in interpretive training, but instead it is the fourth-most-common textbook used by North American post-secondary interpretation courses (Lackey, 2008). On page 278, Knudson, Cable, and Beck write, 'An interpreter of Native American heritage with proper, authentic attire rivets listener attention and gives stronger credence to the story of a historic battle than do signs, slide shows, or museum exhibits' (1995, p. 278). This is tokenizing and racist. If the point is simply that personal interpretation is generally more engaging than non-personal interpretation, that should be said, and Indigeneity should be left out of it. One cannot ignore this, nor the authors' failure to address the complexity and ethics of the research needed to inform interpretation of Indigenous heritage and perspectives.

Existing training for heritage interpreters working within protected areas in settler-colonial nations inadequately prepares students for their careers. Existing training and texts do not expressly address research, knowledge construction, or power relations. In places built on the genocide of Indigenous peoples and in parks where heritage necessarily involves both settlers and Indigenous peoples, these gaps are distressing. The solution to this problem is more than simply 'hire more Indigenous interpreters.' Instead, existing training needs updating and heritage interpretation as a profession needs to become more critical of the information it presents to the public. The remainder of this article first discusses how Indigenous knowledge might be considered in interpretation before offering specific solutions as to how to better integrate Indigenous issues into interpretive practice.

Indigenous and Western knowledge in interpretation

If an interpreter decides to eschew Western science and learn directly from Indigenous communities and Elders (as they should), they must navigate two different epistemologies. Western knowledge derives from a positivist, linear, reductionist epistemology. In this tradition, knowledge is sought by examining small parts to understand larger systems and by understanding interactions as a series of easily observable cause/effect relationships.

Indigenous knowledge, meanwhile, is holistic and focuses on relationships between knowledge, people, and all of Creation (the 'natural' world as well as the spiritual). Indigenous knowledge is the process of participating fully and responsibly in such relationships, rather than specifically the knowledge gained from such experiences (McGregor, 2004). As referenced above, McGregor teaches us that for Indigenous peoples, knowledge is not just about understanding relationships; it *is* the ongoing relationship with Creation.

Interpreters within Indigenous contexts may find themselves turning to Indigenous knowledge to inform their work; it is hard to argue with the appropriateness of asking Indigenous people about themselves rather than turning to non-Indigenous 'experts' for information. However, turning away from pure reliance on Western science and academics and looking toward only Indigenous sources and knowledge presents its own problems, rooted in two areas. First, there is widespread disagreement about what, exactly, Indigenous knowledge is. Second, Indigenous knowledge is community-, culture-, and place-specific. In the context of a transient workforce, this means that employees cannot necessarily apply lessons learned in one season to their work in the next.

The very definition of Indigenous knowledge is subject to debate. Anishinaabe scholar Deborah McGregor (2004, p. 343) flatly states that 'there is no consensus' on a definition of Indigenous knowledge. It is a holistic way of viewing the world, 'not a single concept or construct' (McGregor, 2009, p. 74) and is an entire way of relating to Creation. Further, one of the defining characteristics of Indigenous knowledge is that it is 'an integral part of an Indigenous culture' (Doberstein & Devin, 2004, p. 49). Different communities and nations hold different Indigenous knowledge and one must be careful not to divorce Indigenous knowledge from the community that produced it.

Put another way, 'Indigenous knowledge is not another frontier for science to discover' (Wavey, 1993). This applies as much to interpreters as it does to pharmaceutical companies seeking the next wonder drug in the Amazon. Indigenous knowledge is part and parcel of the community in which it originates; it is not something to be learned quickly and extracted from its context.

Given that interpretation relies heavily on a seasonal workforce, this is an important point. It would be deeply inappropriate, for example, for an interpreter to offer a program focusing on 'Indigenous knowledge about the desert' at Grand Canyon National Park (USA) in one season and to offer the same program at Zion National Park (USA) the following season. Attempting to bridge Western and Indigenous knowledge systems is a difficult task. Below, I offer some suggestions for addressing these roadblocks, in addition to ways to address the broader problems this article considers.

Walking along the Western-Indigenous boundary

To this point, we have been considering the challenges interpreters face when working in settler-colonial and Indigenous contexts. We have seen the problems associated with relying heavily on Western academia and 'the experts' as Tilden would have one do. Similarly, there are challenges to navigate if one decides to eschew Western researchers in favor of Indigenous knowledge and sourcing.

But all hope is not lost. There are ways to address these problems; interpreters should not give up and simply choose not to approach Indigenous content and themes in their work for fear of doing it incorrectly. Rather, interpreters should embrace the opportunities that come with interpreting Indigeneity while using the challenges to grow professionally. This final section will provide concrete advice for interpreters working in Indigenous contexts.

I believe there are two broad approaches towards addressing these challenges. First, interpreters must improve how they work with Indigenous communities. Second, interpretation needs to develop within itself a culture of self-criticality and reflection. Such a culture is currently lacking in the field and is urgently needed.

How can interpreters engage with Indigenous peoples? First, for programs/media that will incorporate Indigenous knowledge, the community which holds that knowledge must have a substantial and substantive role in the entire interpretive process. It is inappropriate to seek approval for an already-designed exhibit or program. Indigenous communities and Elders should not be vetting completed interpretive media/programs but should instead be equal partners its design and creation.

However, consulting directly with Indigenous knowledge-holders and/or relying on them alone to present Indigenous-focused interpretation is not always possible. For example, a seasonal interpreter at Big Bend National Park (USA) will likely not have the time or the physical ability to visit with Apache or Comanche Elders while developing programs at the start of the summer. Therefore, permanent employees should have primary responsibility for ensuring interpretation of Indigenous themes is conducted appropriately.

This may very well mean that seasonal staff are precluded from developing Indigenous-focused programming or media. The interpretation of Indigeneity occurs against a violently imperial historical context, particularly within parks created and maintained at the expense of Indigenous nations, and it is the responsibility of the profession to ensure it does not contribute to this imperialism but instead affirms Indigenous sovereignty and rights. This can be partially accomplished if permanent employees drive Indigenous-focused interpretation, instead of allowing the task to fall seasonal employees who may lack the time and resources needed to do so.

Finally, interpreters need to be listening to what Indigenous people say about research in their communities, for interpreters are consumers and communicators of research. As Tilden, Beck and Cable, and others have elegantly articulated, interpreters are the link between specialists and the public. It is a moral and ethical imperative for interpreters to carefully consider the effects of their sources (i.e. the research interpreters draw upon) on Indigenous people. To neglect to do so only serves to support and perpetuate improper research practices that reinforce old patterns of Indigenous people as 'studied, exotic "others"' (Kovach, 2010, p. 170).

Interpretation needs to look inward to solve these problems. This means both being attuned to the political nature of interpretation – that one's program and treatment of Indigenous themes will influence visitors' conceptions of Indigeneity – and understanding one's role to be as much a researcher as a communicator. This duality, as researcher-communicator means interpreters should not only question their sources, but they should question themselves:

- Are they being a good steward of Indigenous culture, history, and perspectives?
- If they are working with a specific community and are seeking to reveal a particularly sensitive issue/topic, do they have the community's assent to do so?
- How will this interpretive offering concretely aid in the remediation of centuries of colonialism?
- Why have they, the interpreter, chosen to tell this story, to give this program, and/or to forge this visitor-resource connection?

If the answer to this final question is some form of 'because it will advance Indigenous interests by ____,' it is likely that the interpreter is on the right track.

Interpreters need to be explicitly taught to interrogate their research and sources. As Māori scholar Graham Smith has pointed out, 'Every thesis is laden with the author's own interests that are often submerged in the text' (interview with Kovach, 2010, p. 90). Interpreters must be critical of their sources. They cannot unquestioningly rely on someone else to declare 'the facts.'

Some useful questions to ask of one's sources include:

- How was the knowledge informing a program or exhibit gathered?
- Was the research conducted by a researcher who parachuted into a community, extracted knowledge, and used it in a manner that harmed rather than helped the community?
- Or was the knowledge gathered by a researcher whose work was informed by and responded to the community's needs and interests?
- Why was this research done? Was it to further the community's interests? Or the interests of the settler-colonial state?
- Who performed this research?
- Did the researcher(s) have the social license from the community to conduct this work?
- If they interviewed people, did the interviewees have credibility within the community and to speak on its behalf?

Reflecting on both oneself and one's sources and research is crucial. It can help one to understand how one's 'beliefs shape ... what we take as "truth" and knowledge' (Kovach, 2010, p. 111). An explicit consideration of biases, positionality, and of the subjectivity inherent in filtering research, deciding what is important or useful, discarding the rest, and presenting those selections to the public is urgently needed in interpretive practice.

The above questions may also help interpreters consider their conceptual frameworks, which 'make visible the way we see the world' and 'provide insight into a researcher's beliefs about knowledge production' (Kovach, 2010, p. 41). This is a crucial point, and one that interpreters need to fully understand. Interpreters are gatekeepers, adjudicating knowledge and privileging some of it over other knowledge. Interpreters are much more than communicators and interpretation itself is more than an art. It is a powerful exercise of power that can erase or advance particular understandings of the resource being interpreted. One's conceptual framework shapes how one wields this power.

Yet, the power of interpretation is underappreciated and under-discussed. To reiterate an earlier point, most of the interpretation-focused scholarship discusses program methods, outcomes, and/or evaluation. The field, including those who instruct aspiring and new interpreters, needs to become more attuned to the other aspects of interpretation. This begins with understanding interpreters' roles as researchers and knowledge-arbiters. Recasting interpreters as researchers and communicators underscores the need for permanent employees and supervisors to appropriately guide seasonal employees whose time and ability to conduct ethical, appropriate research for programs is limited.

As Kovach says, professional 'responsibility implies knowledge and action. It seeks to genuinely serve others and is inseparable from respect and reciprocity' (p. 178). A responsible interpreter working with Indigenous themes will understand the methods and relations underpinning the research she is drawing on for her programs and media. She will ensure that, if she cannot consult with Elders about her program, the work informing it is congruent with Indigenous peoples' expectations for ethical research.

A responsible interpreter considers why she is presenting a program or exhibit and honors the stories shared with and by her by ensuring that they advance Indigenous interests, rights, and sovereignty. A responsible interpreter does not begin with the facts as they have been determined, as Tilden instructs. A responsible interpreter instead begins with thinking about *how* those facts have been determined.

Conclusion

Indigenous themes are interwoven with the fabric of North America and Australia (for example). It is short-sighted and inappropriate to consign thinking about interpretation of Indigenous themes and the issues raised in this article to only the staff at overtly 'Indigenous' sites, for these continents are entirely the traditional territory of Indigenous peoples. It is incumbent upon all interpreters to be proficient in their chosen profession. Understanding how their job fits into broader historical and political contexts (i.e. colonization) is a part of that proficiency. Likewise, learning to be an interpreter needs to involve much more than discussions of theme construction or proper exhibit layout. It needs to also include a deep appreciation of the interpreter as a researcher who critically analyzes sources and questions them rather than simply turning to 'the experts' and accepting what they say at face value.

Interpretation is a profound exercise of power; interpreters are burdened with the responsibility to ensure that their programs are accountable to and do not harm the resources, cultures, and peoples being interpreted. In an Indigenous context, as I have demonstrated here, this responsibility demands an understanding of the role of the Western academy in settler-colonial imperialism, an appreciation for the differences between Western and Indigenous knowledge, engaging in critical self-reflection, and committing to interpreting Indigenous themes in a manner that serves Indigenous interests. In contexts without apparent Indigenous links, it means ensuring that the interpreter is not un/knowingly contributing to the attempted erasure of Indigenous people and culture in settler-colonial societies.

Note

1. I will use "park" and "protected area" interchangeably in this paper. I recognize that distinctions exist between these; my choice is purely stylistic to ensure varied word use.

Acknowledgements

The author thanks the two anonymous reviewers for their time and comments. Ravi de Costa, Deborah McGregor, and Paul Wilkinson have his gratitude for their guidance and support. He lives on the traditional territory of the Anishinaabe Nation, the Haudenosaunee Confederacy, the Huron-Wendat, and Métis peoples. He acknowledges the myriad ways he benefits from their oppression. Toronto continues to be home to many Indigenous peoples, including the current treaty-holders, the Mississaugas of the New Credit First Nation.

Disclosure statement

No potential conflict of interest was reported by the author.

ORCID

Chance Finegan http://orcid.org/0000-0001-5063-8959

References

Ashmore, W. (2004). Social archaeologies of landscape. In L. Meskell & R. W. Preucel (Eds.), *A companion to social archaeology* (pp. 255–271). Malden, MA: Blackwell.

Bacher, K., Baltrus, A., Barrie, B., Bliss, K., Cardea, D., Chandler, L., & Lacome, B. (2007). *Foundational competencies for all NPS interpreters.* Washington, DC. Retrieved from https://www.nps.gov/idp/interp/101/FoundationsCurriculum.pdf

Beck, L., & Cable, T. (2002). *Interpretation for the 21st century: Fifteen guiding principles for interpreting nature and culture.* Urbana, IL: Sagamore.

Bewell, A. (2004). *Natures in translation: Romanticism and colonial natural history.* Baltimore, MD: Johns Hopkins University Press.

Brochu, L., & Merriman, T. (2002). *Personal interpretation.* Fort Collins, CO: InterpPress.

Clark, B., & Scheiber, L. (2008). A sloping land: An introduction to archaeological landscapes on the high plains. In B. Clark, & L. Scheiber (Eds.), *Archaeological landscapes on the high plains* (pp. 1–16). Boulder: University Press of Colorado.

David, B., & Thomas, J. (2016). Landscape archaeology: Introduction. In B. David, & J. Thomas (Eds.), *Handbook of landscape archaeology* (pp. 27–43). Walnut Creek, CA: Left Coast Press.

Disko, S., & Tugendhat, H. (2014). *World heritage sites and Indigenous peoples' rights.* Copenhagen: IWGIA, Forest Peoples Programme, and the Gundjeihmi Aboriginal Corporation.

Doberstein, B., & Devin, S. (2004). Traditional ecological knowledge in parks management: A Canadian perspective. *Environments, 32*(1), 47–69.

Foxlee, J. (2007). Cultural landscape interpretation: The case of the sorry rock story at Uluru-Kata Tjuta National Park. *Tourism Recreation Research, 32*(3), 49–56.

Geniusz, W. D. (2009). *Our knowledge is not primitive: Decolonizing botanical anishinaabe teachings.* Syracuse: Syracuse University Press.

Gilmore, K. (2008). Ritual landscapes, population, and changing sense of place during the late prehistoric transition in eastern Colorado. In L. Scheiber & B. Clark (Eds.), *Archaeological landscapes on the high plains* (pp. 71–114). Boulder: University Press of Colorado.

Griffiths, T., & Robin, L. (1997). *Ecology and empire: Environmental history of settler societies.* Seattle: University of Washington Press.

Grove, R. (1996). *Green imperialism: Colonial expansion, tropical island Edens and the origins of environmentalism, 1600-1860.* Cambridge: Cambridge University Press.

Ham, S. (1992). *Environmental interpretation: A practical guide for people with big ideas and small budgets.* Golden, CO: North American Press.

Keller, R. H., & Turek, M. F. (1998). *American Indians & national parks.* Tucson: University of Arizona Press.

Knudson, D., Cable, T., & Beck, L. (1995). *Interpretation of cultural and natural resources.* State College, PA: Venture.

Kovach, M. (2010). *Indigenous methodologies: Characteristics, conversations, and contexts.* Toronto: University of Toronto Press.

Lackey, B. (2008). The state of interpretation in academia. *Journal of Interpretation Research, 12*(1), 27–38.

LaDuke, W. (2015). *All our relations: Native struggles for land and life* (2nd ed.). Chicago, IL: Haymarket Books.

Lambert, L. (2014). *Researching for Indigenous survival.* Lincoln, NE: University of Nebraska Press.

Madin, E. M. P., & Fenton, D. M. (2004). Environmental interpretation in the Great Barrier Reef Marine Park: An assessment of programme effectiveness. *Journal of Sustainable Tourism, 12*(2), 121–137.

McGregor, D. (2004). Coming full circle: Indigenous knowledge, environment, and our future. *The American Indian Quarterly, 28*(3), 385–410.

McGregor, D. (2009). Linking traditional knowledge and environmental practice in Ontario. *Journal of Canadian Studies. Revue D'etudes Canadiennes, 43*(3), 69–100.

Moscardo, G. (2007). Interpretation, culture and the creation of place. *Tourism Recreation Research, 32*(3), 57–64.

Nadasdy, P. (1999). The politics of TEK: Power and the "integration" of knowledge. *Arctic Anthropology, 36*(1–2), 1–18.

Nadasdy, P. (2005). Transcending the debate over the ecologically noble Indian: Indigenous peoples and environmentalism. *Ethnohistory, 52*(2), 291–331.

National Association for Interpretation, Brochu, L., & Merriman, T. (2012). *Certified interpretive guide training workbook.* Fort Collins, CO: InterpPress/National Association for Interpretation.

Nelson, G. (2016, August 22). Why has the National Park Service gotten whiter? *High Country News.* Paonia, CO. Retrieved from http://www.hcn.org/issues/48.14/why-has-the-national-park-service-gotten-whiter

Ogle, B. (2016). Evaluation of new interpretive elements in an aquarium renovation. *Journal of Interpretation Research, 21*(1), 41–44.

Powell, R. B., & Ham, S. H. (2008). Can ecotourism interpretation really lead to pro-conservation knowledge, attitudes and behaviour? Evidence from the Galapagos Islands. *Journal of Sustainable Tourism, 16*(4), 467–489.

Pratt, M. L. (2007). *Imperial eyes: Travel writing and transculturation.* New York, NY: Routledge.

Ren, Q., & Folta, E. (2016). Evaluating environmental interpretation with mixed method: A case study at the International Crane Foundation, Wisconsin. *Journal of Interpretation Research, 21*(2), 15–19.

Smith, L. T. (1999). *Decolonizing methodologies.* New York, NY: Zed Books.

Staiff, R., Bushell, R., & Kennedy, P. (2002). Interpretation in national parks: Some critical questions. *Journal of Sustainable Tourism, 10*(2), 97–113.

Tilden, F. (2009). *Interpreting our heritage.* Chapel Hill, NC: UNC Press.

Tubb, K. N. (2003). An evaluation of the effectiveness of interpretation within Dartmoor National Park in reaching the goals of sustainable tourism development. *Journal of Sustainable Tourism, 11*(6), 476–498.

Tuck, E., & Gaztambide-Fernandez, R. A. (2013). Curriculum, replacement, and settler futurity. *Journal of Curriculum Theorizing, 29*(1), 72–90.

Tuck, E., & Yang, K. W. (2012). Decolonization is not a metaphor. *Decolonization: Indigeneity, Education, and Society, 1*(1), 1–40.

Veracini, L. (2010). *Settler colonialism: A theoretical overview.* London: Palgrave Macmillan.

Veracini, L. (2011). Introducing settler colonial studies. *Settler Colonial Studies, 1*(1), 1–12.

Wall, K. (2011). "A sliver of the true fort": Imagining Fort Edmonton, 1911–2011. *Journal of Heritage Tourism, 6*(2), 109–128.

Wavey, R. (1993). International workshop on Indigenous knowledge and community based resource management: Keynote address. In J. Inglis (Ed.), *Traditional ecological knowledge: Concepts and cases* (pp. 11–16). Ottawa: International Program on Traditional Knowledge and International Development Research Centre.

Wolfe, P. (2006). Settler colonialism and the elimination of the native. *Journal of Genocide Research, 8*(4), 387–409.

Index